SACRAMENTO PUBLIC LIBRARY
5814
4/2010
D0457602

Also by Marcelo Gleiser

The Dancing Universe: From Creation Myths to the Big Bang

The Prophet and the Astronomer: Apocalyptic Science and the End of the World

A Tear at the Edge of Creation

A RADICAL
NEW VISION
FOR LIFE IN AN
IMPERFECT UNIVERSE

Marcelo Gleiser

FREE PRESS
New York London Toronto Sydney

FREE PRESS
A Division of Simon & Schuster, Inc.
1230 Avenue of the Americas
New York, NY 10020

Kepler's Mysterium Cosmographicum printed from the original with permission from Herzog August Bibliothek Wolfenbuettel: 40 Astron.

First Free Press hardcover edition April 2010

FREE PRESS and colophon are trademarks of Simon & Schuster, Inc.

DESIGNED BY ERICH HOBBING

Manufactured in the United States of America

10 9 8 7 6 5 4 3 2 1

Library of Congress Cataloging-in-Publication Data
Gleiser, Marcelo.
A tear at the edge of creation : a radical new vision for life in an imperfect universe / Marcelo Gleiser.—1st Free Press hardcover ed.
p. cm.
Includes bibliographical references and index.
1. Cosmology. 2. Life (Biology) I. Title.
QB981.G575 2010
523.1—dc22

2009046247

ISBN 978-1-4391-0832-1
ISBN 978-1-4391-2786-5 (ebook)

In memoriam

Carl Sagan (1934–96)

Your absence is most evident

The Universe is asymmetric and I am persuaded that life, as it is known to us, is a direct result of the asymmetry of the Universe or of its indirect consequences.

—Louis Pasteur

It [is] hopelessly narrow-minded . . . to imagine that all significant laws of physics had been discovered at the moment our generation began contemplating the problem. There would be a twenty-first-century physics and a twenty-second-century physics, and even a Fourth-Millennium physics.

—Carl Sagan, *Contact*

Pure thought didn't supersede creative engagement with phenomena as a way of understanding the world twenty years ago, hasn't in the meantime, and won't anytime soon.

—Frank Wilczek, Summary talk at "Expectations of a Final Theory," Cambridge University, September 2005

I don't want to discourage string theorists, but maybe the world is what we've always known: the Standard Model and general relativity.

—Steven Weinberg, CERN Courier, September 2009

CONTENTS

Introduction xiii

Part I | Oneness

1. Burst! 3
2. Fear of Darkness 9
3. Transition 12
4. Belief 16
5. Oneness: Beginnings 20
6. The Pythagorean Myth 22
7. Living the Platonic Dream 25
8. God, the Sun 29
9. To Hold the Key to the Cosmos in Your Mind . . . 33
10. Kepler's Mistake 36

Part II | The Asymmetry of Time

11. The Big Bang Confirmed 43
12. The World in a Grain of Sand 45
13. Light Acts in Mysterious Ways 47
14. The Imperfection of Electromagnetism 51
15. The Birth of Atoms 54
16. From Creation Myths to the Quantum: A Brief History 58
17. Leap of Faith 64
18. The Jitterbug Cosmos 67
19. The Universe That We See 72
20. The Faltering Big Bang Model 78

21. Back to the Beginning 80
22. Exotic Primordial Matter 83
23. A Small Patch of Weirdness 87
24. Darkness Falls 91
25. Darkness Rules 95

Part III | The Asymmetry of Matter

26. Symmetry and Beauty 101
27. A More Intimate Look at Symmetry 105
28. Energy Flows, Matter Dances 109
29. Violation of a Beautiful Symmetry 113
30. The Material World 117
31. Science of the Gaps 123
32. Symmetries and Asymmetries of Matter 129
33. The Origin of Matter in the Universe 135
34. A Universe in Transition 140
35. Unification: A Critique 147

Part IV | The Asymmetry of Life

36. Life! 155
37. The Spark of Life 159
38. Life from No Life: First Steps 164
39. First Life: The "When" Question 167
40. First Life: The "Where" Question 171
41. First Life: The "How" Question 177
42. First Life: The Building Blocks 182
43. The Man Who Killed the Life Force 187
44. *L'Univers Est Dissymétrique!* 190
45. The Chirality of Life 194
46. From So Asymmetric a Beginning . . . 199
47. We Are All Mutants 206

PART V | THE ASYMMETRY OF EXISTENCE

48. Fear of Darkness II 215
49. Is the Universe Conscious? 216
50. Meaning and Awe 222
51. Beyond Symmetry and Unification 224
52. Marilyn Monroe's Mole and the Fallacy of a Cosmos "Just Right" for Life 228
53. Rare Earth, Rare Life? 235
54. Us and Them 241
55. Cosmic Loneliness 246
56. A New Directive for Humanity 248

Epilogue: Garden of Delights 253
Notes 257
Bibliography 267
Acknowledgments 271
Index 273

INTRODUCTION

> If we don't have a distinctive position or velocity or acceleration, or a separate origin from the other plants and animals, then at least, maybe, we are the smartest beings in the entire universe. And that is our uniqueness.
>
> —Carl Sagan (1985)

> All philosophy is based on two things only: curiosity and poor eyesight . . . The trouble is we want to know more than we can see.
>
> —Bernard le Bovier de Fontenelle (1686)

Sometimes high walls must be torn down to reveal new vistas. For millennia, shamans and philosophers, believers and nonbelievers, artists and scientists, have tried to make sense of existence, searching for ultimate explanations of reality. Central to this quest for meaning is the notion of Oneness, which suggests that all that exists is somehow interconnected. Many religions call for a deity that transcends the constraints of space and time, a being of absolute power who designed the world and who controls, to a greater or lesser extent, the fate of humanity. Every day, billions of people go to temples, churches, mosques, and synagogues to pray to their divine incarnation of Oneness. Not too far from the houses of worship, scientists at universities and laboratories try to put together explanations of the natural world based on a surprisingly similar notion: under the apparent complexity of Nature, there is a simpler underlying reality where everything is somehow interconnected. In this book, I will argue that belief in a physical theory that unifies the secrets of the material world—a *hidden code of Nature*—is the scientific equivalent to the religious belief in Oneness. We could call it "monotheistic science." Some of the greatest scientists of all time, Kepler, Newton, Faraday, Einstein, Heisenberg, and Schrödinger, believed and searched for this elusive code. Nowadays, theoretical physicists, especially those trying to understand questions related to the composition of matter and the origin of the universe, call this code the "Theory of Everything" or the "Final Theory." Is such a quest justified? Or is it fundamentally misguided?

Fifteen years ago, I would never have guessed that one day I would be writing this book. A true believer in unification, I spent my Ph.D. years, and many more, searching for a theory of Nature that reflected the belief that all is one. The most popular candidate for such a theory was (and still is) called *superstring theory*. This is a proposal whereby the fundamental entities of matter are not point particles like electrons, but wiggling strings of submicroscopic size that live in nine spatial dimensions. The mathematical elegance of the theory is compelling, as is its promise to fulfill the age-old dream of unification. Many of the brightest minds in theoretical physics are working to advance this theory and some of its rival proposals.

The cornerstone of any unification theory is the notion that a more profound description of Nature possesses a higher level of mathematical symmetry. Echoing the teachings of Pythagoras and Plato, this idea carries with it an implicit aesthetic judgment that such theories are more beautiful, and, as the poet John Keats wrote in 1819, that "beauty is truth." And yet, as we investigate the experimental evidence for unification, or even for how such ideas can be experimentally verified, we find very little hard data supporting them. Of course, symmetry remains an essential tool in the physical sciences. But during the past fifty years, discoveries in experimental physics have shown time and again that our expectations of higher symmetry are more expectations than realities.

Although at first very distressing at a personal level, this realization eventually led my work in a new direction. I began to recognize that it is not symmetry but the presence of *asymmetry* that best represents some of the most basic aspects of Nature. Symmetry may have its appeal, but it is inherently stale: some kind of imbalance is behind every transformation. As I explain in this book, from the origin of matter to the origin of life, the emergence of structure depends fundamentally on the existence of asymmetries.

Slowly, my thoughts converged into an aesthetic based on imperfection rather than perfection. I found that asymmetry is beautiful precisely for being imperfect just as Marilyn Monroe's mole is beautiful. The revolution in modern art and music started more than a century ago is, to a large extent, an expression of this aesthetic. Now, it's time for science to let go of the old aesthetic that espouses perfection as beauty and beauty as truth.

This new take on science has far-reaching implications. If we are here because Nature is imperfect, how common is life in the universe? Can we guarantee that, given similar conditions, life will emerge elsewhere? What about intelligent life? Are there other thinking beings in the cosmos? Quite unexpectedly, my scientific quest led me to a new understanding of being human: science turned existential.

Behind the age-old search for Oneness is the belief that life cannot be an accident, that our existence must be planned to be meaningful. Whether we were created by gods, as many religions profess, or are the fruits of a universe geared for life, our presence here *must* have a reason. Otherwise, we would be left with a depressing alternative: a meaningless life in a purposeless universe. Many are offended by the notion that we are here merely due to a series of accidents. Why should we be able to understand so much, to love and to suffer, to create works of inexpressible beauty, only to perish and, with very few exceptions, to be forgotten within a few generations? Why should we be able to sense the passage of time, if we cannot some how master it? No, we must either be god-like creatures or part of some cosmic master plan.

Well, what if we *are* an accident, a rare, precious accident, animated aggregates of atoms capable of self-awareness? Should we think less of humanity for not being part of a grand plan of Creation? Should we think less of the universe if there isn't a hidden code of Nature, a set of fundamental laws that explains all there is? I will argue that we should not. On the contrary, the revelations of modern science, while justifying the view that, indeed, there is no grand plan for Creation, put humans in a very central role. We can even call it the dawn of a new "humancentrism." We may not be "the measure of all things," as the Greek philosopher Protagoras proposed around 450 B.C.E., but we are the things that can measure. As long as we keep wondering about who we are and about the world around us, our existence will have meaning.

Let's consider this point in more detail. After only four hundred years of modern science, we have created a remarkable body of knowledge that stretches all the way from the inner confines of atomic nuclei to galaxies billions of light-years away. As we have peered with our amazing tools into the realm of the very small and of the very large, we have uncovered worlds within worlds of unsuspected richness. At every step along the way, Nature has surprised and enchanted us and will continue to do so. As we pieced together a narrative of how the universe evolved from a hot primordial soup of elementary particles to create increasingly complex material structures, we have marveled at the endless diversity of forms to be found everywhere. Most mysterious of all, we still wonder how inanimate matter came to life, and how living bags of molecules evolved to turn a rocky planet into a crucible of biological activity.

Seeing the richness of life here, and knowing that the laws of physics and chemistry apply across the universe, we turned our eyes to our planetary neighbors, eagerly searching for companionship. Sadly, in spite of our hopes, we have found only barren worlds. Beautiful, yes, but destitute of any obvious signs of life. Even if some form of life is hidden in the Martian underground, or in the dark subsurface oceans of Jupiter's moon, Europa, it will surely not resemble anything like sentient beings, pondering as we do the meaning and purpose of life. If such beings do exist—the search is on for extraterrestrial intelligence—they will be so remote as to be, for all practical purposes (and leaving aside wild speculations), inexistent to us. As long as we remain alone, *we* are how the universe reflects upon itself: our mind is the cosmic mind. This revelation has profound consequences. Even if we are not the creation of gods or of a purposeful cosmos, we are here and we can reflect upon existence.

Our living planet floats precariously in a hostile cosmos. We are precious because we are rare. Our cosmic loneliness should not incite despair. Instead, it should incite our will to act, and act fast, to protect what we have. Life on Earth will continue without us. But we can't continue without Earth, at least not for very long. And time is a luxury we don't have.

• • •

A note to my readers: This book was written for anyone interested in how science's remarkable discoveries influence our worldview and help shape our culture. Whenever possible, I used analogies and metaphors to illustrate scientific concepts. There are no formulas or equations of any sort. Technical terms are carefully avoided and, when used, explained as they are introduced. However, since the text deals with cutting-edge ideas in cosmology, particle physics, biology, and astrobiology, at times the reading may feel intense. If that happens to you, don't get discouraged. Skip the paragraph, or even the chapter, and move on. The book is divided in five parts. Everyone should start with Part I, "Oneness." If you initially are somewhat reluctant to dive into the science, you may skip to Part V, "The Asymmetry of Existence." I hope you will then go back and read Parts II, III, and IV to fill in the blanks. They present the beautiful science that attempts to describe the origin of the universe, the origin of matter, and the origin of life, respectively, emphasizing the role of asymmetries and imperfections in each: from the multiverse to the Big Bang; from the Big Bang to atoms; from atoms to cells; from cells to humans; and from humans to extraterrestrial life. I also provide a bibliography for those who would like to complement their reading.

PART I

Oneness

1 | BURST!

There were no witnesses to what was about to happen. "Happen" didn't yet exist. Reality was timeless. Space also didn't exist. The distance between two points was immeasurable. The points themselves could be anywhere, hovering and bouncing. Infinity tangled into itself. There was no here and now. Only Being.

Suddenly, a trembling, a vibration, an ordering began. Like roiling waves, space shuddered and swelled. What was near became far. What was now became past. As space and time were born, change began to happen: from Being to Becoming. Space bubbled; time unfurled. Soon, matter coalesced from the joint heaving of space and time, seeping out of its pores. This was no mundane substance: nothing like us; nothing like atoms.

This matter stretched space, made it inflate like a swelling balloon. This balloon became our Universe.

This is the creation story of our generation. The Holy Trinity here is Space, Time, and Matter. There is no Creator, no divine hand to guide the unfolding of the cosmos from Being to Becoming, from a timeless to an evolving state. The Universe happened on its own, a bubble of space that burst into existence from a sea of nothingness: *creatio ex nihilo*, creation out of nothing. That's hard for us to fathom, as everything that we see happening seems to have a cause behind it. Should the Universe be any different? Could it really have emerged from nothing? Without a cause?

The first link in the long chain of causation from cosmic birth to now, the cause that started it all, is known traditionally as the

First Cause. To do its job—trigger creation—it must necessarily be uncaused. The challenge, of course, is how to implement this mysterious, common-sense-violating, uncaused First Cause. Is science up to the task? Religions mostly use gods to bypass the creation dilemma. That works well for them, since physical laws and common sense don't apply to gods. Being immortal, they are indifferent to causation: they exist, supernaturally, beyond time. In the book of Genesis, all-powerful, eternal God manipulated "nothingness" with words, and there was light. For Jewish, Christian, and Muslim believers, He is the First Cause. All comes from Him, and He comes from nothing. It then follows that since He is perfect, what He creates must be perfect, too. Until, that is, Adam and Eve ate from the Tree of Knowledge and changed everything: curiosity and desire expelled us from Paradise, made us less than godly. Since then, as mere mortals, we've been aching to reconnect with what we've lost, to become one with God's perfect creation. This noble-sounding quest has led us astray for too long. We need a fresh start.

According to some modern theories concerned with the origin of space, time, and matter, there exists a quantum nothingness, a bubbly foam of prototype universes called the "multiverse" or the "megaverse." A few current theories state that the multiverse is eternal and hence uncaused. Occasionally, from the cosmic froth bubbles of space spring forth—baby universes. Some grow, while most shrink back to the nothingness whence they came. A clever interplay between gravity and matter allows baby universes to be born with zero energy cost: creation out of nothing. Time starts ticking when a bubble bursts into existence and begins to evolve, that is, when there is change to be accounted for. Multiverse theories propose that we live in one of these growing bubbles, one that emerged as randomly as a particle that is shot out of a radioactive atomic nucleus. Our bubble, our Universe with a capital *U* (to differentiate the portion of the universe we can actually measure from either hypothetical model universes or portions of the universe beyond our current measurements), has the apparently rare distinction of having existed long enough for galaxies, stars, and people to emerge: we result from the random birth of a highly improbable long-lived cosmos that has grown complex enough to spawn creatures capable of pondering their origins. This is a far cry from the premeditated supernatural creation portrayed in Genesis. But does it fully address the question of how everything came to be?

Clever as this scientific version of creation is in its attempt to do away with the First Cause, it still must be formulated according to accepted physical principles and laws: energy must be conserved; the speed of light and other fundamental constants of Nature must have the right values to ensure the viability of our Universe. Furthermore, a quantum nothingness, with its bubbly soup of prototype universes, is not quite what we would call the absence of everything. The point is, we humans cannot create something out of nothing. We need the materials; we need the joining rules. This limitation of ours is clearest when we try to make sense of the first of all creations, that of the Universe. Do not let claims to the contrary fool you, even if they involve awe-inspiring terms like *quantum vacuum decay, string landscape, extra-dimensional space-time,* or *multi-brane collision:* we are far from having a convincing, empirically validated (meaning tested or even testable) scientific narrative of creation. Even if, one day, we do devise such a theory, it must always be qualified as a *scientific* theory of creation, based on a series of assumptions.

Science needs a framework, a scaffolding of principles and laws, to operate. It cannot explain everything, because it needs to start with something. This something must be taken for granted. Examples of such starting points are the axioms of mathematical theorems—unproved statements accepted as being self-evident and thus supposedly true—and, in physical theories, a number of laws of Nature, such as energy and electric charge conservation, that often are extrapolated to be valid well beyond their tested range. Seeing how well these laws work for the natural phenomena we can observe and measure, we assume that they held in the extreme environment prevalent near the Big Bang, the event that marked the beginning of time. But we cannot be sure—and scientists should never claim they are—until there is clear experimental confirmation. "Extraordinary claims need extraordinary evidence," noted University of California paleontologist J. William Schopf.

On the other hand, modern cosmological theories do explain the physical processes that took place very close to the beginning of time, an achievement that is—and should be loudly advertised as—truly amazing. We can now state with confidence that the Universe did spring from a hot and dense soup of elementary particles a little under 14 billion years ago, even though we still don't know how the spring sprang. We know that the young, minute-old cosmos forged the lightest chemical elements, and that exploding stars forged—and continue to forge—the heavier ones needed for life. We understand the workings of the genetic code and the mechanism behind the staggering variety of animals and plants on Earth. Barring the existence of other self-aware beings capable of theorizing about life and death, we—imperfect accidents of creation—are how the Universe thinks about itself. To my mind, this is a life-transforming revelation, the substance of this book. Even though we live in no special place in the cosmos and play no starring role in the grand scheme of things, the fact that we carry this banner—alone or not—does make us very special. For this very reason, we must be extra careful. In spite of all our achievements, we will do well to remember that our story is just our story, imperfect and limited as we are; we will do well to remember not to go after absolute truth but after understanding. As Tom Stoppard reminds us in his play *Arcadia*, it's not knowing everything, but wanting to know that matters.

Wonderful as it is, science is a human construction, a narrative we create to make sense of the world around us. The "truths" that we obtain, such as Newton's universal law of gravitation or Einstein's special theory of relativity, are indeed impressive, but always of limited validity. There is always more to explain beyond the reach of a theory. New scientific revolutions are going to happen. Worldviews will shift. Yet, vain as we are, we place too much weight on our achievements. Our successes have led us to believe that these partial truths are scattered pieces of a single puzzle, the components of a Final Truth, waiting to be discovered. Great minds of the distant and recent past have devoted decades of their lives in search of this Holy Grail, Nature's hidden code: Pythagoras, Aristotle, Kepler, Einstein, Planck, Schrödinger, Heisenberg. The list is long. Thousands more are doing so today. Knowingly or unknowingly, they are heirs to a philosophical tradition rooted in ancient Greece that links perfection and beauty with truth. Over the centuries, this tradition was fused with monotheistic belief: God's creation was perfect and beautiful. To understand it, to search for immortal truth, became the highest of aspirations. Since the birth of modern science in the early

1600s, a passion akin to religious fervor has led to the widespread conviction that the puzzle can be solved, that we are closer than ever, that Nature's hidden code will soon be unveiled in all its glory. British physicist Stephen Hawking, as many before him, metaphorically compared the achievement to "knowing the mind of God." But is that so? Are we truly getting any closer? Or are we lost, searching for the unattainable? Should we instead be asking why we *need* to believe so badly in this Final Truth? Should we be asking why we are so convinced that it is there to be discovered? Does the experimental and observational evidence at hand truly point this way? Or is this Final Truth simply the scientific incarnation of the monotheistic tradition of the West, a yearning for a God that reason exorcised from spiritual life?

Given that the Final Truth necessarily explains the origin of the Universe, we now see how these two quests are one and the same: the Final Truth contains the First Cause; the First Cause contains the Final Truth. Can we, limited beings that we are, explain creation in all of its astonishing complexity?

We know at least two answers:

"Sure!" exclaim the Unifiers. "There is a fundamental set of physical laws, writ deep into Nature's essence, behind all there is. Given time, we will uncover these laws and make sense of it all. Together, these laws are the embodiment of the unified field theory, the supreme expression of the hidden mathematical symmetry of Nature. We call it the Theory of Everything."

"Sure!" exclaim the Believers. "We already know all the answers. They are written in our Holy Book. Creation is the work of our all-powerful God. Only a supernatural power could exist before space. Only a supernatural power could be before time. Only a supernatural power could transcend material reality to create it."

Are we limited to these two choices? Is there a third alternative? For millennia, we have lived under the mythic spell of the One. Kneeling at our temples or searching for the mathematical "mind of God," we have yearned for a connection with what is beyond the merely human; we have dreamt of an abstract perfection that we could not find in our lives. In doing so, we closed our eyes to ourselves, refusing to accept the fragility of our existence. It is now time to move on. It is now time to shake free of the old imperative for perfection and embrace the lessons of a new scientific worldview that explores the creative power of Nature's imperfections and accepts that there are limits to knowledge.

The journey will be humbling, as we face the smallness of our existence in a vast, indifferent cosmos. And yet, small that we are, our very existence makes us unique. Thinking aggregates of inanimate atoms, we are rare and precious. In a few millennia, we have achieved the power to change the course of our planet's history and, with it, our own. Humanity is at a crossroads. The decisions we make now will shape our future and that of our planet. It is time to understand that preserving life is what really matters.

2 FEAR OF DARKNESS

When I was a boy, I was terrified of the dark. What the eyes couldn't see, the mind would invent. There was a big closet in my room, made of jacaranda, a now-rare noble wood from the Brazilian tropical forest. Full of patterns, the wood would come alive at dusk, moving and contorting in impossible ways. The night-light at the foot of my bed only made things worse, its pale green flicker animating the dance of wooden shapes. Like a human ostrich, I would bury myself under the covers and put a pillow over my head, hoping that if I couldn't see the shadow-beings, they couldn't see me either.

But the fear persisted. Did something touch my foot? What was that strange creaking noise? I could feel air whooshing by my exposed nose. "They" were getting nearer. Disaster was sure to happen any moment now . . . They would strip my sheets and pillows, and their fangs would dive deep into my neck, draining my life away. If I were to survive, I would have to fight. Imbued with a faint sense of heroism, I would peek out from under my pillow, trying to convince myself that there was no one there, that it was all my imagination. That's what my father kept telling me, over and over. "If you're so afraid of the dark, why do you watch all these horror movies? Why are you reading all these comics with stories of vampires and werewolves? What's *wrong* with you?"

He only knew half the story. At the age of ten, my time was filled with horror stories and the supernatural. Fear became an addiction. I didn't simply watch the movies and read the books. I *was* a vampire, or was, at least, about to become one. True, an apartment in tropical Copacabana was no decrepit castle in Transylvania. But it was clearly meant to happen. I even had the canines capable of making needle-sharp punctures on a sheet of paper to prove it.

"Obviously a bizarre psychosomatic effect," my exasperated father, a Harvard-trained dentist, would protest.

By the time I was eleven, my morbid attachment had become more intellectual. I would ride the bus to the National Library in downtown Rio to research vampirism and read some of the classics. Becoming a vampire was the only way I could be half dead, both dead and alive. It was the only way I could become immortal. To the uninitiated, traditional vampires are dead during the day, when they lie in their secluded coffins, and alive after dusk, when they prowl the shadows in search of human blood, the secret of their immortality. Was there a creature more towering than Count Dracula, the Prince of Darkness, capable of defeating death, of controlling humans—especially beautiful women, with his hypnotic powers, capable of flying as a bat and dematerializing as fog?

There was a reason for my preteenage morbidity. In one word: *loss*. When I was six, my mother died in tragic circumstances. She was thirty-eight years old. Now that I have children of my own, I realize how devastating such an early loss can be. It's not just the obvious fact that you suddenly become the child without a mommy in the playground, the one the other kids are sorry for. "Poor Marcelo, he doesn't have a mommy like you do . . . you should go play with him." How many times have I overheard well-intentioned mothers and nannies say this to their children? It's not just the humiliation of being different or the pain of lacking the deep physical and emotional attachment to the woman who gave birth to you. The most painful part of not having a mother is not having a mother; it's not having someone to hold you when you are scared, to praise you when you come home with good grades or after winning a game, someone you know will always love you, no matter what. I saw my friends walking out of school holding hands with their moms, all smiles and hugs, and I felt cursed. The worst part of not having a mother is knowing that she will not see you grow, that she will never be part of your life again; knowing that there will be an empty spot at your graduation, at your marriage, at the birth of your first child. It's the absence that hurts. The worst part of losing your mother is that it is forever.

I couldn't accept this. I had to transcend the boundaries of time, the world of the living, and find a way to bring her back to me, or to go to her. I had to see her again, feel her warm skin, look into her bright brown eyes, hear her laugh. All I remembered were the tears, the sadness, the sobbing. If only I could control time, I could change things. If only I could control life and death, I would be with her again.

To my young, impressionable mind, incapable of differentiating between reality and fantasy, this jump into the world of the supernatural, of vampires and other creatures that could defy death, seemed an obvious choice. During the day at school, I heard stories about God and the Old Testament, of how He brought down the Flood and drowned all of humankind but for one family, of how He turned sticks into serpents and waters to blood, of how angels would come down from Heaven to wrestle with mere mortals. Then there was Rabbi Loew of seventeenth-century Prague, who gave life to a huge clay man after inscribing magic words onto his forehead. If that was what I was learning in school, was it so absurd to believe in other supernatural beings? How could the school psychologist call me disturbed, when in class we learned that God could turn people into salt statues, and, in the Catholic school down the road, even resurrection was allowed?

3 TRANSITION

I entered my teenage years in a sort of trance. There were many instances when I could swear my mother's ghost hovered at the end of our apartment's long corridor, in a flowing white nightgown. Her face seemed to carry all the sadness of the world. I soon became convinced that her apparition was trying to tell me something. Whether she was truly there or in my mind, the vision, and the feelings it awoke in me, were very real. And gradually, I understood what she was trying to tell me. It was not that I should embrace death to be near her. Instead I should embrace life; I should celebrate her memory by living all that she didn't; I should make her proud and happy to be my mother. Because the truth is, alive or dead, she would always be my mother. When you don't have a mother, you invent one, so as to fill the enormous emotional abyss. And this is true not only when losing a parent. I chose to tell this particular story because I live it. *Any* loss leaves a gap that must be filled. The question is, of course, how do we fill it?

The transition to life had begun. I became a serious volleyball player. I started classical guitar lessons. I focused on schoolwork and girls. By the time I reached fourteen my mother's ghost stopped appearing, a sign I took to mean that she was at peace, knowing I had chosen a life-embracing path.

At about this time, I discovered science. I had known it before, of course, mostly from the dull, dispassionate presentations of my elementary and junior high school teachers. In spite of their best efforts to make it seem uninteresting, however, I was fascinated. I gawked at the TV along with everyone else when Neil Armstrong and Buzz Aldrin planted the U.S. flag on the Moon. The destructive power of the H-bomb both terrified and mystified me, as did the possibility that, for the first time in history, we could obliterate civilization at the push of a few buttons. Stanley Kubrick's *2001: A Space Odyssey,* in full Cinerama, deeply impressed me with its mix of science and mysticism. Were there greater intelligences living somewhere in the vastness of space? Could *they* have been our creators? Were they watching us from an invisible distance, indistinguishable from gods? My father's much-treasured copy of Erich von Däniken's compelling and utterly absurd *Chariots of the Gods* threw more wood into my adolescent fire. There it was, *proof* that aliens had been here before and yes, that they were immensely smarter than we. Still, though I went through a phase of believing that the impressive Nazca Lines in Peru were landing lanes for sophisticated alien spaceships, or that the pyramids of Egypt were built under alien tutelage, a skeptical little voice in my head persisted in asking inconvenient questions: Why were they only interested in us in the distant past, when we had such primitive technology? Why weren't they back, now that we had started to crawl into outer space, to give us a much-needed push to the stars?

This exposure to sensationalist science, allied with the typical teenage onslaught of testosterone, transformed my fear of darkness into a love of the night and its mysteries. I switched from a Dracula wannabe to a Victorian scientist wannabe. After all, even Dr. Van Helsing, the vampire slayer, was a respected professor at a European university who used reason and knowledge to destroy evil. Mary Shelley's *Frankenstein* was not a horror movie, but a sci-fi novel exploring the cutting-edge research of its time, the power of electricity to make muscles twitch and, possibly, bring the dead back to life.[1]

There was magic in science, I began to realize, a kind of magic all the more powerful for being real, for being the creation of living humans and not of imaginary supernatural entities: the magic of uncovering Nature's deepest secrets. I became deeply suspicious of religion and its tales. Worse, I became a cynic, seeing how many believers have killed and still killed in the name of their gods. What kind of religious morality would not just condone, but spur on, the murder of innocents? As a bumper sticker I saw recently put it, "Do you really think Jesus would own a gun?" And then there was the problem of suffering. Where was God when my mother died? Why me? Was I a sinner? Were my older brothers or my father sinners? Where was He when I prayed and asked for help and got nothing in return? What of the horrible disasters that punctuate humanity's history? Earthquakes and volcanic eruptions burying whole cities; tsunamis; hurricanes; the monstrous killings of men by other men, the Holocaust we studied so much in school, the Stalinist and Maoist purges of millions, and too many other genocides to list. Arguments such as "God moves in mysterious ways" or that "God has better things to do than respond to a little squirt's prayers" or that "men's affairs are men's affairs and not God's" sounded to me like a cop-out. It became clear to me that if God had something to do with the origin of the world and of life, He obviously had lost interest in His creation. There had to be another way to find meaning.

I started to devour popular science books written by venerable authors like the sci-fi icon Isaac Asimov and George Gamow—Gamow being the creative brain behind the Big Bang model—and, of course, Albert Einstein and Leopold Infeld's *The Evolution of Physics*. Of the many things I learned, one hit me the strongest: if we are to understand the Universe's deepest secrets—and that is the grand goal of the physical sciences—we have to search for the hidden symmetries of Nature. There must be a rational order behind it all, accessible to human reason. Scientists seemed to believe that the mathematical expression of this order, encoded in the symmetries of natural phenomena, is the truest expression of beauty.

The enchanting notion of an underlying order in the world resonated deeply with my needs. It had a calming, comforting effect. If, on the face of it, life is chaos, don't despair: look deeper, and you will find order and wisdom. The great German astronomer Johannes Kepler expressed it well when he wrote in 1629, a year before his death, "When the storm rages and the shipwreck of the state threatens, we can do nothing more noble than lower the anchor of our peaceful studies into the ground of eternity." Eternal truth must be coded into Nature's hidden mysteries. I vowed to lower the anchor of my peaceful studies next to Kepler's and search for the timeless rational essence of reality. I understood that the search for eternal truths, as it transcends the frailties of human life, renders loss meaningless.

I began to see science as a heroic pursuit. High-minded men and women sharing a common goal, exchanging knowledge, unveiling Nature's innermost secrets, following in the footsteps of the great sages of antiquity. The specifics were even more awe-inspiring: the theory of relativity and its interpretation of space and time into a bendable, four-dimensional space-time; black holes and the mysteries of time travel; atoms and their awesome power to create and destroy; life and its unknown origins; and, last but not least, the question of questions, that of the origin of the Universe itself. What could be more life-affirming, more stimulating, than devoting myself to this search? Like the fabled heroes of countless sagas, I was ready to embark on this pilgrimage, ready to be transformed by the search. The temple doors were open, and the solutions to the deepest mysteries of existence were inside, waiting to be discovered.

My path lay open ahead of me. I would passionately devote myself to the study of math and physics, become a scientist, and embark on the search for eternal truth, striving to unearth Nature's secrets. Science was a rational connection to a reality beyond our senses. There was a bridge to the mysterious, and it did not have to cross over supernatural lands. This was the greatest realization of my life. I was ready to become a theoretical physicist—and not just *any* kind of theoretical physicist: I was ready to become a full-blown Unifier, a pursuer of Nature's hidden code. I became a believer.

4 BELIEF

There is tremendous power in belief. Denying it is foolish. Those convinced that in an ever more technological society, secularism is the unavoidable future should take a good look around. In June 2008, the Pew Forum for Religion and Public Life released the results of one of the most comprehensive polls on religious faith ever undertaken in the United States, involving more than thirty-five thousand individuals eighteen years and older.[2] To the question "Do you believe in God or a universal spirit?" 92 percent answered "yes." Of those, 71 percent responded that they were *absolutely* certain, and 21 percent, while believing in something, were unable to pin down the nature of their belief. Remarkably, only 5 percent answered they do not believe. The final 3 percent refused to answer. The margin of error quoted was one percentage point. That is, of every ten people you meet in the United States, on average seven are absolutely certain God exists, even if there are variations on what *God* means.

These results make the United States one of the most religious nations in the world. Even if the belief in God is less prevalent in most countries in Europe and East Asia, there is no question that we live in a world where the concept of a supernatural deity is very present. The enormous scientific advances of the past four hundred years have not translated into radically different numbers of believers when compared to those living in ancient Greece or Egypt. Even if the percentage of believers in the time of the pharaoh Akhenaten (circa 1350 B.C.E.) was, say, 99.9 percent, that's only a 7.9 percent difference from the number for the modern United States. In fact, discounting a handful of countries (including most of the Nordic European countries, the Czech Republic, France, Vietnam, and Japan), all other countries in the world count fewer than 50 percent of atheists and agnostics among their population.[3] These numbers say something very important about ourselves: our need to believe trumps all evidence against that belief. In other words, *wanting* to believe that something exists goes a long way toward being convinced of that something's existence, whatever that something is. In the same Pew poll, 49 percent of the respondents claimed that their prayers are answered at least a few times a year: for one in every two Americans, the communication with a supernatural deity is not only possible, but also effective.

During the past few years, a number of books have addressed the so-called war between science and religion from a fresh perspective. Scientists such as Richard Dawkins and Sam Harris, philosopher Daniel Dennett, and British journalist and polemicist Christopher Hitchens, a group sometimes referred to as "the Four Horsemen," have taken the offensive, deeming religious belief a form of "delusion," a dangerous kind of collective madness that has wreaked havoc upon the world for millennia. Their rhetoric is the emblem of a militant radical atheism, a view I believe is as inflammatory and intolerant as that of the religious fundamentalists they criticize.*

This radical approach only furthers entrenchment and bitterness. Extremism is a very inefficient diplomat, as a quick glance at the history of religions teaches us. To call religious people ignorant, insane, or simply stupid may feel good but completely misses the point. Let us leave aside possible social and psychological benefits of religion, such as refuge for the needy, a sense of identity and community, and emotional guidance in the face of loss and bereavement, all of which could presumably be obtained through secular pathways. Still, there is a fundamental reason why people cling to their faith-based beliefs, though they can't be empirically validated. In fact, empirical validation has nothing to do with the enduring power of religious faith: in general, the more mysterious the creed, the more ardent the belief. Most people believe in the supernatural because they cannot accept the finality of death. They are afraid of being forgotten, of reverting to nothingness, of losing their loved ones. They think of the billions

* Dawkins defends himself from accusations of fundamentalism by saying that contrary to religious extremists, he would easily change his mind if shown the evidence. I would imagine that an orthodox rabbi or a mullah would also convert to Christianity if Jesus appeared in his living room floating on a rainbow. But maybe I'm an optimist and he would say it was the devil tempting him.

of people who have passed by this Earth, rich and poor, kings and slaves, famous and unknown, people who, like them, have loved and were loved, felt joy and pain, and who now are nothing but dust. "Is this it, then?" they wonder. "Do we live, love, struggle, suffer, just to be forgotten a few generations later? If all we have is a few years, and not necessarily always joyful ones, what is the point of trying so hard? Is life pointless?"

With more questions than answers, people choose to believe, clinging to a faith that promises to lift them beyond the confines of matter and time. To ridicule this basic human need is to betray an alarming ignorance of what goes on in the hearts and minds of the vast majority of people across the world.

Once, I gave a live radio interview in front of an audience of factory workers from a poor region in central Brazil. As I was going on with my spiel about the Big Bang and the expanding Universe, and how science was ever so close to explaining Creation, a hand went up on the front row. It belonged to a small man, covered in grease, face wrinkled beyond its years. He looked at me with accusatory eyes and yelled: "You, sir, you want to take even God away from us?" I shuddered, knowing too well that his voice is the voice of billions.

Karl Marx famously said that religion is the opiate of the people. If you want to take away religion from the people, you'd better find another very good opium. The high from secular atheism—even with all its appeal to reason and the unavoidable logic of science—is not going to hold sway. At least not as it is usually presented, devoid of any spirituality. To avoid confusion, I should clarify that "spirituality," to me, is not related to a supernatural religious dimension, opposite to the material world. It is also not related to a possible spiritual connection some Unifiers may feel as they pursue their belief in a final theory. What feeds my spiritual connection to Nature is a deeply ingrained feeling that I am, in a very real sense, a physical part of it; that life is a precious gift we should treasure.

Many atheists have said that atheism and agnosticism are not incompatible with spirituality. I couldn't agree more. But creating this compatibility, showing that it is possible, is not easy, especially within the strict materialism of ordinary science. Just saying that Nature is beautiful and that understanding its workings leads to a high level of spiritual fulfillment is not enough. Only a new way of looking at science and at our place in the natural order of things can lead to a religion-free spiritual awakening.

Belief springs from our helplessness in dealing with things we cannot control, predict, or understand. If we are nothing more than flesh and blood, a mere assembly of molecules subject to the laws of Nature, then we have no choice but to follow the course of material things and die, disintegrating into dust. How much more wonderful it is to believe in the afterlife, in nonmaterial entities capable of bypassing the rigid limitations imposed by materialistic reasoning! If science is to help us, in the words of the late Carl Sagan, as a "candle in the dark," it will have to be seen in a new light. The first step in this direction is to admit that science has its limitations, as do the scientists who do it. This way, science will be humanized. We should confess to our confusion and sense of being lost as we confront a Universe that seems to grow more mysterious the more we study it; we should be humble in our claims, knowing how often we must correct them. We should, of course, share the joy of discovery and the importance of doubt. Perhaps more importantly, as I argue in this book, we should explain that there are faith-based myths running deep in science's canon and that scientists, even the great ones, may confuse their expectations of reality with reality itself. We may, as Kepler and Einstein did and too many still do, "dream of a final theory" or "long for the harmonies," to use the titles of two excellent science books by Nobel laureates Steven Weinberg and Frank Wilczek (with co-author and wife, Betsy Devine). But we should not turn a blind eye if Nature seems to have other plans. And the concrete evidence we have today indicates that it has.

5 ONENESS: BEGINNINGS

The idea of Oneness, the idea that underlying the perceived diversity of the world, there is a simpler, all-encompassing reality, can be traced back to monotheistic belief: there is one God, and this God created all that is.* If one believes that everything derives from God or from His/Her transcendent nature, it then follows that all is one: everything that exists and will exist comes from a single source, and to that source it will return. Creation, in all its splendor and misery, in all the beauty and ugliness of its myriad forms, is how God manifests His presence in time. Creation *is* God in time.

Oneness has been with us for millennia. It was no coincidence that I invoked pharaoh Akhenaten above. The first recorded mention of monotheism is from his reign, circa 1350 B.C.E. More specifically, from himself, in his *Great Hymn to Aten,* where he wrote,

> O sole god, like whom there is no other!
> Thou didst create the world according to thy desire . . .

and other exhortations to the single deity Aten. Akhenaten declared himself the only conduit between humans and God, and ordered the destruction of statues and imagery of previous gods, condemning

*Even in polytheistic faiths, there usually is an alpha-god. For the ancient Greeks, Zeus is master of Mount Olympus, all-powerful ruler of the gods. For the Hindus, Brahman is the all-pervading essence of all things, the Divine Ground of time, matter, and space, the creator and destroyer of all existence. In some versions of Buddhism, allegedly a religion without a God, the figure of the Buddha acquires a superhuman, transcendent nature that survived the physical death of Siddharta Gautama, its founder. In Mahayana Buddhism, the central concept of Dharmakaya represents the eternal aspect of all things, the truest essence of the Universe. Variations aside, here we are mostly concerned with the conception of a central God, which is present in one way or another in all major world religions.

his ancestors' faith as pagan. Intolerance toward other religions was already present at the early stages of monotheistic belief: in choosing one God, you discriminate against all others. Although there is controversy among scholars as to whether Akhenaten's influence persisted after his death, Sigmund Freud raised an interesting possibility in his book *Moses and Monotheism*. Freud claimed that Moses was actually an Atenist priest forced to leave Egypt with his followers after Akhenaten's death. According to this thesis, while the monotheistic pharaoh failed, his faithful priest succeeded magnificently, even if with a little help from God himself, as the Good Book tells us.

Without going into the historical merit of the Freudian claim, it is certain that religious ideas did circulate and cross-fertilize throughout the Middle East. Jordan's Pella Migdol Temple records the ascendancy of monotheism farther north at about the same time.[4] With the increase in trading due to better ships and ground routes, people traveled more, exchanged views, and learned from each other. As a result, monotheistic ideas spread westward across the Mediterranean. Around 600 B.C.E. in Greece, they started to transition from religion into philosophy. As we will see next, it is here that Oneness takes hold of the intellectual West.

6 THE PYTHAGOREAN MYTH

Thales, from the Turkish coastal town of Miletus, is considered the father of Western philosophy. As with most of the pre-Socratic philosophers—those remarkably creative men who lived before or around the time of Socrates (circa 469–399 B.C.E.), little to nothing is known of his life and work. Even so, oral tradition and texts written centuries after his death, especially those by Aristotle and the Greek historian Diogenes Laërtius (circa 200 C.E.), attribute to Thales the distinction of having made the first scientific statement about the world: "All things are made of a single substance." Thales believed in a unifying material principle behind all that exists. In Nature's choreography of eternal creation and destruction, everything that exists springs forth from this material and reverts back to it. Right from the very start, Greek natural philosophy is bound to Oneness.

To Thales, water was the material essence of the world, no doubt due to its ability to constantly change and transform while keeping its identity: at the core of his philosophy was the belief that in the transformations of the material world, there is permanence in change. Those who followed Thales, known collectively as the Ionians, kept their master's all-important unifying principle even if they chose different material substances. Anaximenes, for example, proposed air.

Born with the philosophical tradition, the notion of Oneness as Nature's unifying principle would assume many guises and undergo countless variations through the centuries. This "Ionian Enchantment," to quote science historian Gerald Holton's expression for representing the search for unity in science,[5] is as present in modern scientific thought as it was then. More to the point, in his essay "Logical Translation," Isaiah Berlin referred to the search for a unified description of the material world as the "Ionian Fallacy": "A sentence of the form 'Everything consists of . . . ' or 'Everything is . . . ' or 'Nothing is . . . ' unless it is empirical . . . states nothing, since a proposition which cannot be significantly denied or doubted can offer us no information."[6] One of my goals in this book is to unmask the fallacy of the unification enchantment.

A few decades after Thales, Pythagoras combined a form of mathematical mysticism with the Ionian notion of Oneness to create a deeply influential worldview. It's in the Pythagorean legacy that we find the idea that Nature's essence is mathematically symmetrical and thus perfect, a notion that is at the root of the unification dream. For the Pythagoreans, to decipher Nature's secrets was to unveil the symmetries hidden in the deeper layers of reality, below the chaotic diversity of the world. As Plato—who was greatly influenced by Pythagorean thought—argued, the world that we see and hear is a distortion. It is only through thought and ideas that we can find the true essence of reality. This essence is written in the language of mathematical forms and their relations. The only perfect circle is the idea of the circle that exists in your mind. Bertrand Russell, the prominent philosopher and mathematician who won the 1950 Nobel Prize in literature, wrote in his classic *History of Western Philosophy* (1946), "Pythagoras . . . was intellectually one of the most important men that ever lived, both when he was wise and when he was unwise."

Modern scholars have argued that Pythagoras quite likely never proved the famous theorem that bears his name or, for that matter, developed the tools of mathematical proof.[7] Much of what is attributed to him is either the work of his followers or an elaborate fabrication of Plato's disciples Speusippus and Xenocrates, who invoked Pythagoras's legendary ancient authority to support the more mathematical aspects of their mentor's philosophy. The Pythagorean fiction spiraled further in the hands of Plotinus and other Neoplatonists of the early Middle Ages, and again during the Renaissance; all of them were eager to establish a link between mathematics and man's mystical experience of God.

Be that as it may, since antiquity the Pythagorean myth has nurtured the dreams of those who seek Nature's hidden code. Instead of ascribing a specific kind of matter at the core of all reality, as did the Ionians, the Pythagoreans believed that *numbers* were the key to Nature's essence.

Once you believe that Creation is the work of a rational God, mathematics becomes the key to unlocking its secrets and to establishing a union with the Creator. The mythic Pythagoras was the one man who had achieved this union, a demigod capable of superhuman feats, the philosopher-saint we should aspire to become. Discoveries attributed to him, such as his theorem and the relationship between harmonic sounds and integer numbers, were direct insights into the mind of God. Only this Greek master could hear the harmony of the spheres, the consonant humming that planets produce as they zoom about the Earth in their proportionate circular celestial orbits. Pythagoras (or his followers) proposed that the same mathematical ratios that hold between integer numbers and harmonic sounds also hold between the distances to the celestial orbs. So, two guitar strings, one twice the length of the other and thus in a 2:1 ratio, will produce consonant sounds a harmonious octave apart when struck together. Likewise, Saturn is roughly twice as distant as Jupiter, and hence their distances also stand in a 2:1 ratio.*

The Pythagoreans proposed that numbers unify Nature, and that human reason, with its remarkable ability to understand and interpret relationships between numbers, can decipher Nature's hidden code. Given that the power of a myth is not in its being right or wrong but in its being believed, it is the legacy of the Pythagorean myth rather then the real work Pythagoras did or didn't that interests us. In the collective imagination, the Pythagorean mathematical mysticism became the bridge between human reason and divine intelligence. During the late Renaissance, it inspired the work of the man who turned the cosmos upside down.

*In modern numbers, the average distance between Saturn and the Sun is 1,427 million kilometers and that of Jupiter and the Sun is 778 million kilometers, giving a ratio of 1.83:1, not a bad match to a 2:1 ratio. The ancients probably based their numbers on the planets' orbital periods and not on their distances to the Earth, which they couldn't have known. In that case, the result is worse: Saturn with approximately 29 years and Jupiter with 12 years give a ratio of 2.4:1.

7 LIVING THE PLATONIC DREAM

On the day of his death, May 24, 1543, a bedridden, half-paralyzed Nicolaus Copernicus finally saw a freshly printed copy of *On the Revolutions of the Heavenly Spheres,* the magnum opus that summarized his life's work in astronomy. What should have been the happiest of days turned out to be unbearably tragic. For forty years, Copernicus had been quietly convinced that everyone, from the Babylonians to Aristotle, from the great Ptolemy to the inspiring Muslim astronomers who kept the Greek fire burning through the Dark Ages, literally everyone, sage and ignorant alike, had been wrong about the heavens. The beloved onionlike cosmos, where Sun, Moon, planets, and stars revolve around the Earth in neat concentric orbits, was a distortion of reality. Copernicus understood that the Earth was no cosmic center, no hub of Creation: it was just another cosmic wanderer, gyrating with the other planets in earnest devotion about the Sun, source of all light. For four thousand years, the world had been living a lie.

As a result of our Earth-based perspective, we perceive the skies as turning around us. We shouldn't blame the ancients for having thought any differently. Then and now, what we can see and measure determines our perspective on the world. Our imagination may leap forward and expand the possibilities of the real, but ideas will remain mere ideas unless they are confirmed. And since—in spite of our wonderful measuring tools—we will never have complete information about the world, our view of reality will always be limited. We will always be like a fish trapped in a fishbowl, even if our fishbowl grows all the time.

Copernicus knew of a few valiant Greeks, Heraclides of Pontus and Aristarchus of Samos among them, who had proposed alternatives to the Earth-centered model. He also knew that their ideas hadn't held against the Aristotelian onslaught during their time, and that things would only be harder now, after fifteen hundred years of Christian theology had nailed the Earth to the dead center of Creation. It is no wonder that it took Copernicus so long to muster the courage to denounce the world for its folly. The stakes were very high indeed. A different cosmology meant a different worldview; a different worldview meant a different place for man in the cosmos, a different explanation for who we are and our purpose in life. If we were not the center of Creation, were we still God's chosen? If Earth is just a planet, are there people in others? Will they also need to be saved? By our Jesus or their own? Is our Heaven the same as theirs?

To further complicate things, Copernicus didn't have any physics to back him up. True, ordering the planets according to the time they took to complete one revolution about the Sun simplified things enormously and added a sense of aesthetic beauty that was all the rage during these Renaissance days: Mercury, the fast one, had to be the closest to the Sun, as it took only three months to complete an orbit. Venus, with eight months, had to come second. Earth, with one year, had to be third, followed by Mars with two. The giants Jupiter and Saturn took twelve and twenty-nine years, respectively, completing the planetary sextet. But was aesthetics alone a strong enough reason to convince others?

Copernicus was aware that his model was far from perfect. First, his predictions of planetary positions were no better than those achieved by Ptolemy some thirteen centuries earlier with his elaborate choreography of epicycles.* From a practical point of view,

*Not that Copernicus did away with epicycles. They were very much still a feature of his model, as he tried to make circular orbits consistent with astronomical observations. The challenge he and others faced was to mimic the often loopy paths of planets across the skies using only circles. An epicycle is simply an imaginary circle attached to a larger circle. As the epicycle turned, the planet turned along with it, like a person carried along in a Ferris wheel. The added twist is that the person's seat (the epicycle) can also turn. Combining the circular motions of the large circle (the Ferris wheel) and of the epicycle (the person's seat), it is possible to produce elaborate loops (the line traced by the person's head). Ptolemy had used epicycles to obtain predictions of planetary positions with an accuracy equivalent to the size of a full moon in the sky, an amazing achievement.

accurate predictions were what mattered, as they helped boost the quality of astrological forecasting: the better the future positions of the planets were known, the more reliable the forecasting. The same with calendar making, also a challenge at the time. Second, Copernicus didn't offer an alternative to Aristotle's intuitive physics to justify his Sun-centered model: If the Earth were not the center, why should things fall to the ground? Why should the Moon rotate around the Earth but not the Sun and all the other planets? Also, the rotation of the Earth around itself posed problems: If this were true, wouldn't clouds and birds stay behind? Third, observations at the time indicated that unlike the mixings of earth, water, air, and fire that promote the material changes of our world, celestial bodies didn't ever seem to change. Also, they shined with what appeared to be a light of their own. Quite sensibly, Aristotle proposed that they must be made of a different kind of matter, a luminous and everlasting fifth essence not to be found on Earth. Copernicus did offer some criticism of Aristotelian physics in his *On the Revolutions*. However, he didn't propose a new physical framework, but rather relied mostly on the aesthetic appeal of his ideas to justify his new cosmic model.

The lack of both a predictive accuracy better than Ptolemy's and a physical foundation may have contributed to Copernicus's reluctance to make his Sun-centered model public. When he did go ahead, accurate predictions and a new physics, albeit very important, were not his main motivation. Copernicus was intent on living the Platonic dream: he envisioned a cosmos where the planets would cross the skies in circular orbits at constant speeds until Judgment Day, a model of sublime simplicity and beauty. What other shape but the perfectly symmetric circle could the divine crafter have employed to build the world? How else could the planets parade across the heavens, if not in ordered, steady velocity patterns? Plato's deity, the cosmic architect sometimes called the Demiurge, became Copernicus's Christian God. The cosmos was the embodiment of the divine mind and thus had to reflect its perfection. "[To] deduce the principal thing—namely the shape of the Universe and the unchangeable symmetry of its parts" was Copernicus's self-appointed mission. Since antiquity, no one had attempted to build a more powerful bridge between the aesthetic and religious aspirations of men.

After agonizing for decades, and under constant pressure from his few friends, Copernicus decided that he had to put his proposal forward, even if still imperfect. The details would no doubt be ironed out later, when better observations were available.

The book looked beautiful enough, a ponderous, nicely illustrated volume. However, it came with a nasty surprise. A new preface had been inserted, a text Copernicus hadn't penned, right after his heartfelt dedication to none other than Pope Paul III, where he courageously expressed his view that Scripture should not be used to justify the ordering of the cosmos. To his horror, it was a declaration that the Sun-centered model need not, and should not, correspond to the real world, a negation of his life's work. The words must have cut through his heart like a thousand daggers. They blatantly declared that all key ideas in the book were hypotheses "which need not be true, or even provable." The text was unsigned, giving the impression that Copernicus himself had written it. Only in 1609 Kepler would unmask the real author, the Lutheran theologian Andreas Osiander, who, through a series of mishaps beyond Copernicus's control, oversaw the book's final printing. In spite of Osiander's preface, *On the Revolutions* worked its charm on many of Europe's influential thinkers.[8] Among them was Michael Maestlin, Kepler's teacher of astronomy at the Lutheran university in Tübingen, Germany.

8 GOD, THE SUN

In Peter Shaffer's famous play (and screenplay) *Amadeus*, the clash between mediocrity and genius, between conformism and groundbreaking creativity, reaches tragic consequences when a desperate Antonio Salieri terrorizes sickly Mozart to his death. Salieri's calm control is gradually shattered as he witnesses over and over the immortal beauty of Mozart's music. In a devastating scene, Salieri proudly presents a march to his patron, Austrian archduke and Holy Roman emperor Joseph II. Mozart, portrayed as a wild youth with an exasperating mad laughter, steps in and offers to play the march. To everyone's amazement, Mozart quickly embellishes the trivial melody with his own improvisations, transforming it into a work of art. Salieri's wounded expression mirrors his despair: Why would God bestow so much divine talent to a bumbling simpleton, while to me, his devout servant, he bestowed only oblivion? Mortality and anonymity haunted Salieri's character. Melancholy Maestlin, in his darkest moments, must have fallen victim to the same fears.

In 1589, the year seventeen-year-old Kepler joined Maestlin's astronomy class, *Copernicanism* was a dirty word among Lutheran circles. Earlier, and even more vehemently than the Catholic Church, Martin Luther had denounced the Sun-centered cosmos as pagan foolishness. Maestlin, not one to embrace confrontation, had written an astronomy text where practically no mention of Copernicus is to be found, nor any defense of a Sun-centered astronomy or of a moving Earth. Yet he had measured the motion of the great comet of 1577, establishing—in direct contradiction with the Aristotelian doctrine of immutable skies—that it was located well beyond the sphere of the Moon. As a compromise to his self-imposed intellectual castration, Maestlin would tell his best students about Copernicus and his alternative ideas. Perhaps he secretly hoped that one of them would have the courage he lacked to champion the new worldview. He couldn't have dreamt how far Kepler would take his halfhearted enterprise.

A few years ago, I went to Germany and Prague to research Kepler's life. To my mind, no other scientist in history—not even Galileo and his famous face-off with the Inquisition—embodies so dramatically the archetype of the lone hero struggling to uncover the truth. I wanted to understand the source of Kepler's strength, of his lifelong quest of his faith in the notion of Oneness. I never expected that my search would trigger a deep revision of my own worldview.

Kepler's life was an endless succession of tragedies punctuated here and there by moments of sublime revelation. Son of a mercenary father and a hysterical mother who was almost burned at the stake for witchcraft, he was forced to move from place to place across central Europe, victim of the brutal rivalry between Catholics and Protestants that led to the Thirty Years' War. Plagued by illness and devastating emotional losses, Kepler, the truest of Pythagoreans, embraced the search for cosmic order with an ardor that, in hindsight, can be seen as almost desperate. What life denied him, he would find in the heavens.

The mid-October sun was surprisingly bright when I arrived by train at Weil der Stadt, Kepler's birthplace. A fortified stone wall, punctuated by scattered observation towers, gave the small village the appearance of a child trying to seem tougher than it was. Inside, narrow alleys connected the adjoining houses, each adorned with geometrically arranged beams painted in lively colors, in typical Bavarian fashion. Walking from the train station to my hotel, I could almost feel Kepler's presence. Ignoring the cars and electrical wires, and the teenagers with dyed hair and pierced lips, it was easy to forget this was the twenty-first century. I looked around anxiously, knowing that Kepler had walked these very same streets four hundred years back, when witches were still burned at the Marktplatz, Weil's central square.

Arriving at the charming Hotel Krone Post, I asked for a room facing the central square. When I opened the windows, Kepler was right there, staring at me. His huge statue towers over the square, leaving no doubt that Weil is Keplerstadt, Kepler's town. He sits peacefully, with a knowing look in his eyes, holding a manuscript in his left hand, probably the *New Astronomy*, the book that redefined astronomy, and a pair of compasses in his right. An octagonal pedestal holds the structure, with carved openings along its sides nesting statues of other eminent natural philosophers. Among them is Tycho Brahe, the Danish astronomer-prince whose meticulous observations furnished Kepler with the ammunition to unleash the Copernican revolution with full force. And, of course, there is a spot for Michael Maestlin, Kepler's mentor. While Tycho is portrayed in an arrogant pose, pointing defiantly upward, a long robe covers somber Maestlin, which he clutches closed with his fists as if hiding something underneath. The sculptor couldn't have captured their characters more perfectly.

After circling the statue a few times in reverence, I crossed the square toward Kepler's house, now a museum. Although the original structure burned down in 1648, I was assured that this is a faithful reconstruction. Frau Gnad, the museum caretaker and receptionist, greeted me with enthusiasm. When I explained to her that I was a physicist researching Kepler's life, her eyes lit up. She took me to every room, highlighting the bathtub that Kepler apparently never used (supposedly he took only one bath in his life, which he complained made him ill). Then she showed me a brass model of the Mysterium Cosmographicum.

The Mysterium Cosmographicum was Kepler's remarkable attempt to unify the cosmos under a single framework: a strange concoction of five geometric solids interspersed by spheres aimed to re-create the structure of the solar system. I had seen drawings of it, but never a full three-dimensional rendition. Here was what Kepler believed to be Creation's blueprint, a snapshot of God's mind. Frau Gnad, noticing my trancelike state, discreetly left me to my thoughts.

While still at Tübingen, Kepler embraced the Copernican cause with a fervor that can only be described as devotional. Picking up where Copernicus had left off, Kepler grew convinced that the Sun-centered cosmos was the work of God. As such, it had to be perfectly proportionate, of inhuman beauty. He went as far as relating the new astronomical model to the Holy Trinity: in the center was God, the Sun, beaming the light of Creation in all directions; the outer sphere holding the stars at the cosmic boundary was the Son; and the intervening space linking Father and Son, the medium whereby light diffuses throughout the cosmos, was the Holy Ghost. With brilliant insight, Kepler credited the Sun's light with the power to move the planets in their orbits, calling it a "moving soul" (*anima motrix*). Distant planets, receiving less power, would move slower. Even if light was not the answer, for the first time in the history of astronomy someone proposed that an interaction between Sun and planets explained the motions of the solar system. Forces, and not crystal spheres, held the cosmos together.

Aristotle's Earth-centered cosmos, so vividly invoked in Dante's *Divine Comedy*, had come a long way. Astronomy and theology were ready for a new marriage. Kepler searched furiously for the solution to Copernicus's challenge, the solution to the Mysterium Cosmographicum: "[To] deduce the principal thing—namely the shape of the Universe and the unchangeable symmetry of its parts."

Not surprisingly, Kepler's novel ideas stirred discontent among the conservative Tübingen faculty. Maestlin had lit a fire that quickly spread beyond his control. Conflicted between his secret approval of his pupil, a growing jealousy of his brilliance, and a vowed allegiance to the status quo, Maestlin and his colleagues concocted a scheme to silence the young troublemaker. To Kepler's dismay, a few months before he graduated, he was shipped off to teach mathematics at a Lutheran school in distant Graz. The masters he had held in such high esteem had shattered his lifelong dream of becoming a pastor.[9] Their maneuvering, though, failed dramatically. Forced to rethink his future, Kepler found a new calling. If he couldn't serve God from the pulpit, he would serve him through astronomy, unveiling the hidden code of Nature. As he wrote in a letter to Maestlin, "I wished to be a theologian; for a long time I was troubled, but now see how God is also praised through my work in astronomy."

9 TO HOLD THE KEY TO THE COSMOS IN YOUR MIND . . .

While lecturing on astronomy to a sleepy class, Kepler had the insight that would change his life. During a discussion of the motions of Jupiter and Saturn, he suddenly realized, in true Pythagorean fashion, that the distances between the planets could not be accidental. If God had indeed designed the cosmos, and Kepler had no doubt that He did, everything had to have a rational explanation. Why six planets? Why not three or twenty-five? What sets their relative distances to the Sun?* The answer *had* to be hidden in geometry.

Kepler labored for days, growing increasingly frustrated and desperate. And then, in a flash of revelation, he understood. The structure of the cosmos was indeed determined by geometry. There were only five perfect, or Platonic, solids in three dimensions: the familiar cube (six squares) and pyramid (four equilateral triangles), and the less familiar octahedron (eight equilateral triangles), dodecahedron (twelve pentagons), and icosahedron (twenty equilateral triangles). No other three-dimensional solid made from the same two-dimensional shapes closes on itself. In Kepler's vision, the five solids were nested inside each other as in a three-dimensional puzzle. Between each pair of solids, an imaginary spherical shell would locate the planetary orbits. Five solids allow for six shells, that is, for six planets: Sun in the center—shell (Mercury)—SOLID—shell (Venus)—SOLID—shell (Earth)—SOLID—shell (Mars)—SOLID—shell (Jupiter)—SOLID—shell (Saturn). Furthermore, geometry

* Recall that in Kepler's time Uranus and Neptune were not known.

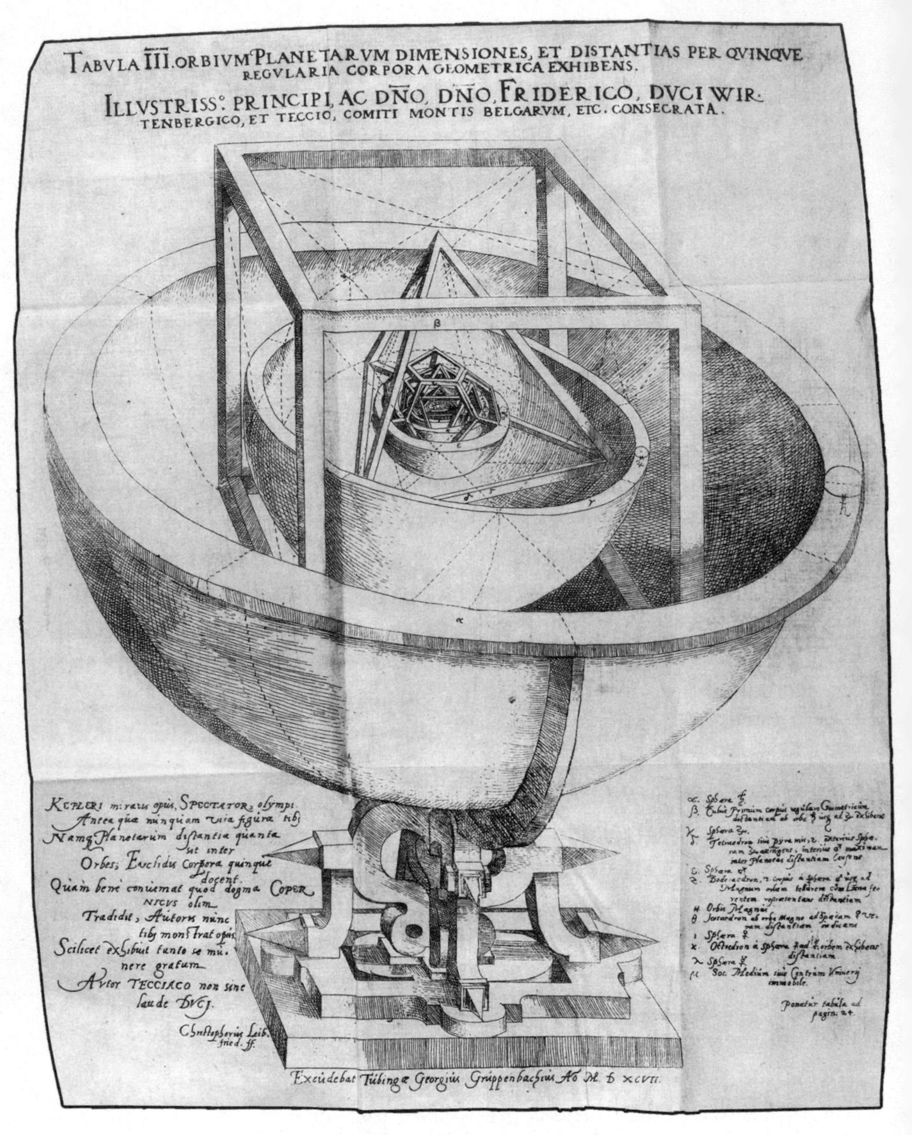

Kepler's Mysterium Cosmographicum

determined the distances between the shells so as to allow the solids to fit neatly. To Kepler's amazement, after fidgeting with the ordering of the solids for a while, he found an arrangement whereby the distances between the solids matched, to a surprising degree, the known distances of the planets.[10]

In a single swoop, Kepler "solved" the greatest astronomical riddle of all time, explaining a priori not only why there are only six planets, but also their distances to the Sun. The seductive power of Kepler's scheme was immense: geometry *uniquely* determines the cosmic blueprint; there is only one solution to the mystery of Creation, and it is the most symmetric possible, as befits God's perfection.

We can only guess at the explosive impact of this discovery in Kepler's psyche. He held the key to the cosmos in his mind . . . At twenty-six, he believed he'd accomplished what no one had before: a glimpse at the inner sanctum of Creation, at the mind of God. Even Maestlin was impressed with Kepler's a priori geometrical solution for the cosmos, and helped his pupil publish it in a book, *The Mysterium Cosmographicum*, which appeared in 1596.

10 KEPLER'S MISTAKE

I took a step back. In front of me was a brilliant man's half-crazed dream of a final theory, an a priori geometric solution to the structure of the cosmos. The order, the symmetry, the exact proportions, all reflected the glory of God's mind. This vision of a geometrical cosmos would stay with Kepler for the rest of his life, even after he changed astronomy forever with his discovery that planetary orbits are not circles, but ellipses. In 1621, Kepler published a new edition of the *Mysterium* with added commentary. Now a mature astronomer, he was as convinced as ever that "when God determined the order of the heavenly bodies," He had in mind "the five regular bodies which have enjoyed such great distinction from the time of Pythagoras and Plato down to our own days." He clung to the Pythagorean dream to the end of his life. More than that, the search for the harmony of the world *was* his life.

How could someone so wrong be so utterly convinced of being right? We have much to learn from Kepler's mistake. In hindsight, it's easy for us to ridicule his creation. After all, there aren't six planets, but eight.* If he could have seen them with the naked eye, he would never have proposed his model, and his career would have taken a different turn. Kepler's blindness was his blessing. He constructed a model of the world with the data he had available. At any given time, including ours, this is the best that anyone can do. What we can measure will always limit our view of reality. Kepler's mistake was to give his vision of reality a finality it didn't deserve. Glimpsing at the hidden code of Nature proved so cathartic that he

* It is a testament to the evolving nature of science that with Pluto's demotion to a "plutoid," a dwarf planet with enough self-gravity to acquire a near-spherical shape, the number of planets recently decreased from nine to eight.

was bewitched and took his belief for the truth. Kepler's mistake was to forget that a final theory is impossible because we will never know all of reality. Then and now, any science that is tainted with blind belief will lead us astray. I looked again at Kepler's creation: a nested finite cosmos, a geometric dream, ordered and precise. At that moment, I knew that my days as a Unifier were over.

How then to explain Kepler's success? The search. Belief in our ideas is a precondition to pursue them. Like the voyager who imagines the Promised Land and charts new territories in his effort to find it, scientists have accomplished much as a result of this pursuit. The vision hangs in the distance, and we do what we can to get to it. Again, Kepler is a perfect illustration of this. Using Tycho's data to improve the accuracy of his polyhedral hypothesis, he discovered the three laws of planetary motion that became the foundation of modern astronomy.* Quoting Holton again, "The search for one grand, architectonic structure . . . is an ancient dream. At its worst, it has sometimes produced authoritarian visions that are as empty of science as their equivalent is dangerous in politics. At its best, it has propelled the drive toward the various grand syntheses that rise above the monotonous landscape of analytical science."[11]

Kepler's synthesis was to bring physical reasoning into astronomy, elevating it from the realm of a mere mapping of the celestial motions. We mentioned that in the *Mysterium,* he suggested that an *anima motrix* emanating from the Sun was responsible for the planetary orbits. This idea was refined in *Astronomia Nova,* where Kepler proposed a magnetic force between the Sun and the planets. These first inklings of gravity will be crucial to Newton's later development. "No other approach," Kepler wrote prophetically, "would succeed than that founded upon the very physical causes of the motions." Even if Kepler's Pythagorean dream of finding a geometric solution to the structure of the cosmos was a religion-driven fantasy, in his search he uncovered the first mathematical laws describing the mechanics of the heavens. It is ironic, but enlightening, that

*The first is that orbits are in general elliptical and not circular. The second is the law of areas, that planetary orbits cover equal areas in equal times, a consequence of their increased acceleration near the Sun due to stronger gravity. The third, which Kepler called the "harmonic law," relates the orbital period of a planet to its average distance to the Sun.

the man who craved symmetry so desperately would be forced to push the circle out of astronomy's center stage. Each planet has its own elliptical orbit, with a bigger or smaller elongation resulting from the planet's detailed formation history: hardly evidence of a perfectly ordered cosmos. Kepler gave us an uglier cosmos but a more accurate science. Imperfection was the price paid for precision, for coming closer to the truth. We now know that Kepler's laws describe the motions not just for this, but also for other solar systems, that is, for planets orbiting other stars. Even if he would have been deeply shocked to know that there are more than six planets in our solar system and thus that his Pythagorean arrangement was false, he would have no doubt rejoiced in learning that his laws are valid across the cosmos. "Those laws," he would have claimed, "are the true signatures of Creation."

After Kepler, the Ionian Enchantment continued to drive the unification dream. During the eighteenth century, largely due to the success of Newton's mechanics, planetary (and many other) motions were seen to reflect general physical principles at work, such as energy and momentum conservation. As a consequence, the search for unity shifted from geometry to the laws of Nature, the body of knowledge that explains how matter organizes itself in different patterns, from the very small to the very large. God became the cosmic lawmaker, and science, the pursuit of His laws.

As I was leaving Kepler's house, Frau Gnad politely approached me. A tremendous coincidence, that night there would be a party in the local high school, the Johannes Kepler Gymnasium, of course, to commemorate the arrival of a new fourteen-inch telescope. Apart from speeches by local authorities, there would be a surprise, which I was sure to enjoy. Would I want to come?

At 7 P.M. promptly, Frau Gnad and a friend picked me up at the hotel. I held on dearly to my life as the black Mercedes zipped around the hills surrounding Weil as if it were the Munich-Frankfurt Autobahn. But we did arrive safely, just a few minutes before the start.

The place was packed: parents, students, teachers, politicians, all celebrating their new window to the sky. It was a formal affair. Unnecessarily long speeches and a trumpet-piano duo playing stiff marches quite distinct from the lyrical harmony of the spheres opened the way to a nice lecture on the solar system and the galaxy. A young woman with a beautiful voice sang—to my surprise—the hit "The Rose," which starts "Some say love, it is a river . . ." as an intro to the pièce de résistance, the surprise Frau Gnad mentioned: a theatrical dialogue between Kepler, an Aristotelian, and a Lutheran theologian, who fiercely debated arguments for and against a Sun-centered cosmos. Kepler, of course, walked out the triumphant but battered winner. Watching the fully costumed actors on the improvised stage, their voices echoing in the silent night, it was impossible not to shed a tear for this great man who longed so deeply for the harmonics.

PART II

The Asymmetry of Time

11 THE BIG BANG CONFIRMED

The year my mother died, 1965, physicists Arno Penzias and Robert Wilson published their findings on what was to become the incontrovertible evidence that our Universe was very hot and dense during the earliest stages of its existence. Being six at the time, I had no clue that as my life took a sharp turn toward darkness, cosmology was entering a golden age of discovery. Penzias and Wilson found that the cosmos is a huge microwave oven with radiation at a temperature of only about 2.73 degrees above absolute zero, that is, of −454.76 degrees Fahrenheit, or −270.42 degrees Celsius. To their great surprise, theorists had predicted that a hot Big Bang would have generated radiation that, after billions of years of cosmic expansion and cooling, would now match precisely the microwaves they had measured with their antenna. In a very real sense, they had found a fossil of the Big Bang.[1]

No other model proposed to date, including the Big Bang's archrival, the steady-state model, which argues for an eternal, unchanging cosmos, can explain the existence of the widespread radiation. The consequences are of mythic proportion: just like us, the Universe has a history, a birth followed by a period of expansion, which is still ongoing. The days of the old, static cosmos of Copernicus and Kepler were long gone. The cosmic expansion implies an arrow of time, a definite sense that as the Universe grows, time moves forward and only forward. Time becomes more than a device to measure change around us; it becomes a cosmic imperative, resolutely pointing toward the future. The implications of this apparently obvious fact are profound. Among them, the asymmetry of time provides a framework to explain the origin of matter and, ultimately, the origin of the stuff of life itself. We are, in a very concrete sense, the product of asymmetries writ deep into Nature's code.

12 THE WORLD IN A GRAIN OF SAND

Among the many mentors I was fortunate to have, I owe an enormous debt of gratitude to my teacher of electromagnetism, Professor Gilson Carneiro, then at the Pontific Catholic University of Rio de Janeiro, where I obtained my physics undergraduate degree. In his wisdom, Professor Carneiro assigned, apart from the usual homework problems and exams, a final project. Students were to present a seminar on a topic of their choice. The only catch was that it had to be related to electromagnetism in one way or another. At the time, 1980, I was already very interested in cosmology. In fact, I had been for quite some time, in a somewhat romantic way, ever since I found out that cosmology deals with the biggest of all mysteries, the origin of everything. "There is this new book by Steven Weinberg," said Professor Carneiro, "called *The First Three Minutes*. It's supposed to be written for the general public and it addresses what you want. Turns out that the whole Universe is immersed in a bath of microwaves, an example of electromagnetic radiation. Read it and give a talk on it."

I was amazed. Weinberg had just gotten the Nobel Prize in physics the year before for his work on, yes, the unification of electromagnetism and the weak nuclear force, two of the four known forces of Nature. What could be more inspiring than having a Nobel-laureate Unifier explain cosmology to you? Weinberg's book affected me deeply. After reading it, there was no question I *had* to work on cosmology and its relationship with unification. The notion that the physics of the very large and of the very small are deeply entangled had a bewitching effect on my late-teen imagination. My mind bubbled with excitement.

A direct consequence of the Big Bang model is that the nascent Universe was denser and hotter. Near the beginning of time, lengths were so tiny and temperatures so high that only the physics of the very small, particle physics, can describe what went on. My father, who loved waxing philosophic when prancing about in his garden, once told me that the macro and the micro were combined in mysterious ways. "You see this grain of sand? The whole Universe is in there!" Was he aware of William Blake's famous lines?

> To see the World in a Grain of Sand
> And a Heaven in a Wild Flower,
> Hold Infinity in the palm of your hand
> And Eternity in an hour.

As with so many questions I had for my father, this one remained unasked. Even as a child, I always sensed that he was right, though I couldn't tell why. Now I knew. The answer was in the interface between cosmology and particle physics. What could be more exciting than connecting the two, the origin of the Universe and the unification of all forces? How could man peer deeper into Nature's hidden code? Weinberg's book had shown me the way: the physics of the early Universe. It was my calling. And I was ready.

13 LIGHT ACTS IN MYSTERIOUS WAYS

All beginnings are humbling. To start climbing the steep mountain toward scientific enlightenment, I had to learn about electromagnetic radiation and those microwaves that bathed the cosmos. In the next few pages I summarize some of the important ideas. They are all part of our present view of the cosmos.

Light, introductory physics textbooks teach us, is an electromagnetic wave visible to the human eye. We also quickly learn that light waves are quite different from the ordinary kind of waves, the ones we see propagating on water or the air compressions created when we speak. These common waves move on something: a material medium supports their propagation. Any wave, light included, is a disturbance that transports energy (and momentum) across space. You throw a stone into a pool, and the energy of the impact is transported away in concentric water waves. Your lungs force air up your throat, and your vocal chords modulate it to create the vibrations eardrums detect as sound.

In 1905, a twenty-six-year-old Einstein suggested that light—or, more generally, any kind of electromagnetic radiation, visible to us or not—is not like other waves. According to the technical expert from Bern's patent office, light doesn't need a material medium to support it. It propagates on its own, in empty space. The suggestion was a shocker. How could this be? Some of the greatest scientists of the nineteenth century had conjectured that light propagated in a notoriously mysterious medium, which, inspired by the Aristotelian filler of Nature's void, was named the luminiferous aether. Its sole function was to provide material support for light waves. A bizarre medium it was indeed: to allow for ultrafast, unimpeded transmission, it had to be millions of times more rigid than steel and yet a fluid; weightless and frictionless to avoid interfering with planetary orbits; and, of course, transparent so that we could see faraway stars. In spite of this very magical combination of physical properties, everyone was confident that the aether was there. It was the only sensible thing for Nature to do.

The overall acceptance of the aether illustrates that it is not only in religion that the will to believe makes the impossible plausible. Things turned ugly when in 1887 Albert Michelson and Edward Morley's experiment failed to find the aether. Not surprisingly, explanations as to why the experiment wouldn't detect the mysterious medium quickly turned up. It *had* to be there. Nature couldn't be this unreasonable. Or could it?

It could. The young patent clerk from Bern was right. There is no luminiferous aether, although it left the stage kicking and screaming. Its demise was a painful lesson to the physicists who, in their blind confidence, demanded that Nature comply with their expectations. Michelson died in 1931 without accepting the consequences of his own experiment. Clearly, one should not reserve the word *delusion* for religious faith. The very important difference, as the aether saga also shows, is that in science the delusion may not be as long-lived as in religion; sooner or later data will come in, theories and models will be cross-examined, and a resolution will come about. Science could not function otherwise. A theory or model that cannot be tested, or that can always be tweaked to escape testing, should not be part of the scientific canon. The luminiferous aether joins the distinguished company of the phlogiston and caloric among the many nonexistent magical substances that scientists have proposed in their attempts to make sense of Nature. We would do well to learn from the devastating crumbling of such certainties and take proposals for strange media and materials with a good dose of skepticism. As we shall soon see, modern astronomical observations suggest that we are again immersed in inscrutable materials. Even if the evidence is compelling, their existence should never be assumed until confirmed.

Eventually, the scientific community accepted that light is an electromagnetic wave capable of propagating in empty space. But what, then, is waving? The answer is electric and magnetic *fields*. Or, more precisely, an electromagnetic field. Picture a small ball of electric charge. The charge creates a field around it, which simply means that other electric charges that come near it will feel its presence; the closer they are to the ball of charge, the stronger they feel it. A field is the spatial manifestation of its source. A hot plate creates a temperature field around it: the closer you are to the plate the hotter it feels. Every electric charge creates (or, equivalently, is the source of) an electric field. Now imagine the ball of charge bouncing up and down like a basketball. Its field bounces with it. In the nineteenth century, physicists realized that bouncing, or more generally, moving electric charges create magnetism. A simple magnet, like the ones used to hang pictures and report cards on fridge doors, owes its magnetism to countless circulating electric charges at the atomic level. The balls of charge here are the negatively charged electrons. Their circulation around the atomic nucleus creates a small magnetic field. To that, we must add their rotating motions about themselves, like tiny tops. The net result, once summed over countless electrons rotating roughly in the same direction, adds up to a collective macroscopic effect: magnetism is electricity in motion.*

In 1831, the great English physicist Michael Faraday discovered that the opposite is also true: moving magnetic fields create electric fields. For example, if you make a circular loop out of a piece of wire and move a bar magnet in and out of it you generate an electric current in the wire. Here the changing magnetic field (from the magnet moving in and out of the loop) creates a changing electric field that, in turn, pushes the charges along the wire. In fact, the magnet's in-and-out motion forces the charges to drift clockwise and counterclockwise along the circular wire loop, creating what is called an alternating current, familiar from modern household appliances. Faraday's discovery revealed a deep relationship between electricity and magnetism: an oscillating electric charge generates changing electric and magnetic fields. As the fields move across space, one bootstraps the other, creating a propagating electromagnetic wave. Entranced by his discovery, Faraday expressed his belief in a deep

*For nonmagnetic materials, the electrons rotate in random directions and the net magnetization, once summed over all the individual contributions, either vanishes or is very small.

unity of Nature: "I have long held an opinion, almost amounting to conviction, in common I believe with many other lovers of natural knowledge, that the various forms under which the forces of matter are made manifest have one common origin." For years he tried to bring gravity into the unification of electricity and magnetism, eventually giving up on it: "Here I end my trials for the present. The results are negative. They do not shake my strong feeling of the existence of a relation between gravity and electricity, though they give no proof that such a relation exists." Faraday was a member of the Sandemanian Church, an orthodox Christian sect with very strict practices. The unity he sought in his science mirrored his monotheistic belief in the God of all Creation. This notion, even if Faraday's Christian God becomes a metaphor for Nature's hidden mathematical order, is the foundation of the modern search for unification.

14 THE IMPERFECTION OF ELECTROMAGNETISM

Electromagnetism is often used as the archetypical example of how two forces, apparently so distinct, can be viewed as manifestations of a single, unified force. According to this view, once we peer deeper into reality, what seems disparate on the surface is revealed as one. In the early 1860s, the Scottish physicist James Clerk Maxwell, in a work of remarkable genius, derived the equations describing all electromagnetic phenomena that had been observed by Faraday and many others. By unlocking the deep mathematical interconnection between electricity and magnetism, Maxwell obtained a groundbreaking result: light is an oscillating electromagnetic wave propagating through space. Its speed in empty space is a mind-boggling three hundred thousand kilometers per second, a number that escapes our comprehension.* In the blink of an eye, light travels seven and a half times around the Earth. We don't know why light travels with this speed and not another. We also don't know why it's always the same: whether its source is moving or standing still, or whether an observer is moving with respect to its source (the same thing, really), the speed of light is always constant. This is why it is a "constant of Nature," a quantity we can measure but not explain, at least for now. What we do know—wild science-fiction speculations aside—is that nothing can travel faster than light. There is no indication that this simple rule has ever been violated. Or that it can be, at least within realistic physical assumptions. The ordering of time depends on it.

Light's constant speed is the cornerstone of Einstein's special the-

* Light slows down when moving through material media, such as water or air. But usually not by much.

ory of relativity. Contrary to popular belief, the special theory of relativity is really based on an absolute. In fact, on two. The constancy of the speed of light is the second postulate of the theory. The first is that the laws of Nature are the same for all observers moving at a constant speed with respect to each other. Science would be impossible if people in a car and in a lab had to deal with different laws of Nature. Ten years later, in his general theory, Einstein generalized this rule to *any* relative motion, not only motion at a constant speed.

In spite of the beautiful relationship between the electric and magnetic forces, there remains a fundamental difference: whereas it is possible to have isolated positive and negative electric charges, such as an electron, no one has ever seen an isolated magnetic charge. Magnets always appear with two "charges," called poles, together. Breaking a bar magnet to isolate the two poles creates two small magnets instead, each with its pair of poles, usually called north and south. If a magnet is broken down all the way to its atoms, we find that each atom is a tiny magnet with its two poles. At the other extreme, the Earth is a giant magnet, with its south and north magnetic poles very close to, but not quite on top of, its geographic poles.*

The absence of "magnetic monopoles," isolated chargelike sources of magnetic fields, has distressed many people. It's a scar on the face of electromagnetic unification, tainting the perfection of their unity: If electric monopoles are so common, why not magnetic ones? How could the two fields be considered truly unified if such obvious disparity between them persists? I remember my disappointment when I first learned of it in Professor Carneiro's class. It felt like a chunk was missing from an otherwise perfect pie. (The class was right before lunch.) A faint feeling of uneasiness settled in, the first bee in my Unifier's bonnet. Could something be wrong with my dreams of a final theory? At the time, I chose to ignore the issue. I still had too much to learn.

* Earth's magnetic field is generated at its core, consisting mostly of a giant rotating sphere of molten iron. The details of the core's rotation are responsible for the direction of the magnetic field and hence of its related poles. The net result is that Earth's magnetic poles are constantly on the move, making an exact alignment with the geographic poles highly unlikely. In fact, the poles have reversed direction hundreds of times over the past billion years and are likely to do so again.

In 1931, the great British physicist Paul Adrian Maurice Dirac proved mathematically that magnetic monopoles are compatible with quantum mechanics, the physics that describes atoms and their constituents. He even showed how the magnetic charge should be "quantized"—appear in integer multiples of a minimum unit—just as electric charge distributions are always multiples of the electron's charge, and amounts in dollars are always multiples of a cent. There is no fundamental law of Nature forbidding the existence of magnetic monopoles, at least that we know of. Yet after almost a century of hunting, magnetic monopoles have refused to show up.* On February 14, 1982, Stanford University physicist Blas Cabrera thought his detector registered the passage of a magnetic monopole, presumably coming from outer space. The physics community reacted with a flurry of excitement. The search intensified. Laboratories across the world turned on their monopole-hunting equipment. Unfortunately, no other monopole was ever detected at Stanford or elsewhere, even with more sensitive apparatus. No one really knows why Cabrera saw what he saw. Quite possibly it was due to a glitch in the detector or a calibration error. Wounding as it may be to our sense of the aesthetic beauty of physical theories, simple magnetic monopoles seem not to exist. Even if they do, they are obviously exceedingly rare. If Nature is telling us that the unification of electricity and magnetism is imperfect, we should listen.

*More sophisticated unified theories, to be discussed in Part III, predict the existence of other kinds of magnetic monopoles. These have not been detected, either.

15 THE BIRTH OF ATOMS

To recap, what we call light, of the luminous kind, is just one type of electromagnetic wave, or radiation. There is a vast spectrum of electromagnetic waves, extending all the way from long-wavelength radio waves to very short-wavelength gamma rays. Visible light is a tiny window in this spectrum, comprising waves with wavelengths of the order of half a millionth of a meter.* We are surrounded by invisible electromagnetic waves. It would be complete chaos if we could see the myriad waves broadcasted by radio stations, cell phone towers, or the infrared radiation emanating from warm objects and people around us. Human eyes (and those of all species) evolved to capture only the most essential visual information the brain needs to create a picture of reality coherent enough to maximize our survival chances. We are creatures of the Sun, and the Sun emits mostly in the visible part of the spectrum (peaking around wavelengths of 500 nm). Appropriately, our eyes are attuned to the dominant type of radiation in our habitat. A lot remains invisible to us, either because the emissions are outside the visible portion of the electromagnetic spectrum, or because their sources are too far or too small. To see Nature in all its glory, we need tools to open the doors of our perception, making us "see" what our eyes can't. Modern astronomy uses telescopes that can "see" distant celestial bodies in all electromagnetic wave types, from radio to infrared to ultraviolet to X-rays to gamma rays. On the other extreme, scientists "see"

* More precisely, the visible eye can sense electromagnetic radiation ranging from 380 to 750 nm, where a nanometer (nm) is one-billionth of a meter, $1 \text{ nm} = 10^{-9}\text{m} = 0.000000001$ meter.

the invisible world of the very small through powerful microscopes and particle accelerators that probe all kinds of minuscule structures, from microbes to atoms and well into the atomic nucleus.

Back to Weinberg's book: its main point is the groundbreaking revelation of modern cosmology, that the Universe as a whole is bathed in electromagnetic waves of the microwave kind, with a dominant contribution from wavelengths around 2 millimeters. Somewhat ironically, the aether, the medium that fills Nature's void, turned out to be light itself! If not visible light, at least one of its invisible long-wavelength cousins. More remarkably, this radiation is a true fossil of a distant past when the cosmos was so hot and dense that no material structures familiar to us existed: no galaxies, no stars, no planets, not even large molecules. All that existed was radiation; the basic constituents of atoms—protons, neutrons, and electrons; and the nuclei of the lightest chemical elements and their isotopic variations (nuclei with same number of protons but different number of neutrons): the hydrogen isotopes deuterium (one proton and one neutron) and tritium (one proton and two neutrons); helium-3 (two protons and one neutron) and helium-4 (two protons and two neutrons); and lithium-7 (three protons and four neutrons). Completing the list were elementary particles called neutrinos, related to radioactivity. We will have much to say about the elusive neutrinos in Part III.*

In the 1960s and '70s we learned that the Universe has a story: like us, it had a birth date and has been growing since then, evolving in time from a very hot and dense initial stage to what it is today, a swelling of cold empty space punctuated here and there with galaxies. The cosmic story describes an increased complexification of matter, starting from a hot primordial soup of its simplest indivisible components, called elementary particles, to their gradual assembly

*The picture is a bit more complicated. Possibly, there were also dark-matter particles present, exotic particles that are as yet unidentified and that are unrelated to ordinary matter. We infer their existence from the way galaxies rotate and from how they move in galactic clusters: galaxies seem to carry much more mass than what is visible in stars and gas clouds. Some of it may be from planets and very dim stars, but measurements indicate that these objects are not enough. The extra mass collects like a halo around the visible galaxy and is believed to be composed of a new kind (or kinds) of matter particles. We will get back to dark matter soon.

into more organized atomic nuclei, atoms, and molecules, eventually leading to stars, plants, animals, and people. The reconstruction of this story, of how matter and cosmos evolved from the simple to the complex, is the central topic of cosmology.

In 1946, the Russian-American enfant terrible of physics, George Gamow, proposed what became known as the Big Bang model. Since the late 1920s, thanks to astronomer Edwin Hubble's remarkable discovery that distant galaxies were moving away from each other at speeds that increased with the distance between them, it was known that the Universe was expanding. A pedagogical pause is needed here. What does it mean, for the Universe to be *expanding*? It's easy to get confused and think of the Big Bang as some kind of explosion and visualize galaxies as little shards of shrapnel flying away from the bang. This image wrongly assumes that space is rigid and that the bits of shrapnel fly away from each other due to the force of the explosion. It also wrongly assumes that there is a center point from which everything originated, the center of the Universe. In cosmology, space is not fixed: it is elastic, capable of stretching and shrinking like a rubber balloon. Hubble showed us that we live in an epoch of cosmic stretching: if you think of galaxies as old-fashioned streetlights placed on the street corners of a large city, the stretching of space *carries* the galaxies away from each other as if the streets themselves were stretching. Furthermore, no point is more important than any other. Everywhere you go, the streets will be stretching and the streetlights will be moving away from you. (If you find these notions strange, don't worry. Physicists struggle with them, too.)

Gamow conjectured that if the Universe is expanding it must have been smaller in the past. Smaller implies denser and hotter, as matter got squeezed into tighter and tighter volumes. Gamow suggested that at earlier times heat and violent collisions would have kept matter in simpler structures. Starting from a time when atoms existed and moving backward, you'd see the dissociation of atoms into free electrons and atomic nuclei. This happened very early on, a mere four hundred thousand years after the Big Bang. Why then? Before that time, the intense pounding of radiation against electrons was effective in stopping them from bonding with protons to form atoms: as a result, atoms couldn't exist then.

Think of it as a love triangle: electrons and protons, with opposite electric charges, are trying desperately to get close and bond, but needy radiation constantly interrupts them by kicking the electrons away. With time, though, radiation gets weaker and unable to avoid the inevitable. It finally departs, leaving protons and electrons to consummate their electric affair. This epoch marks a boundary: before it, no atoms existed, only particles and radiation. After it, the cosmos was filled with atoms and radiation. This radiation, no longer colliding with electrons, cruises freely across space, responding only to the occasional gravitational attraction of large gatherings of matter. Although it was still highly energetic at the time, mostly in the visible and ultraviolet—the Universe was aglow then—it continued to cool off gradually with the cosmic expansion, shifting from the visible to the infrared and, billions of years later, to the microwaves of today. Gamow suggested that this radiation would be a remnant from the age of atomic formation. When Arno Penzias and Robert Wilson detected it in 1965, the Big Bang got the boost it needed. With its main prediction confirmed, it became more than speculative theory. Gamow and his collaborators, Ralph Alpher and Robert Hermann, were vindicated, even if never widely recognized, for their achievements.[2]

With the main idea of the Big Bang model confirmed—that the Universe was hotter, denser, and smaller in the past—an obvious question jumped to the forefront. What happened *really* early on, that is, before the appearance of atoms at 400,000 years? How far back can we turn the clock? Can we go all out, to the beginning of time? Can science finally come to grips with Creation? And if it can, will ideas of unification be vindicated?

16 FROM CREATION MYTHS TO THE QUANTUM: A BRIEF HISTORY

Creation is a loaded word. It means very different things to different people. There is something deeply intimidating about a time before humans, before life itself, a time beyond our control, lost in a past where no Earth or Sun existed, a time before stars.

For millennia, we have made up stories about this faraway time. Our ancestors looked around, saw the way things were, the way Nature operated, and created narratives that tried to make sense of it all, to explain their reality. People of the sea told how the waters came down from heaven, and how the gods separated water from land. Those of the forest told of trees that touched the skies and how the gods sent animals and men to live under their protection. Those of the desert told how their gods fashioned living beings from clay. Among them, the God of the Semites, we are told in Genesis, breathed life into a clay figure, animating it to become the first living man.

Faced with the vastness of Creation, it was clear to these cultures that the transition from a timeless nothingness to the wondrous diversity of the world, from the nonliving to the living, could only have been orchestrated through some awesome power, beyond anything mere humans could handle. This power had to be *super*natural—beyond the natural—so as to be able to shape Nature into the world we are part of.

One of the problems with creation myths is that the gods of different people did different things at different places. There are hundreds of creation stories. The gods of each belief system held sole power over what is and what can be. So deeply ingrained was this belief, so life defining, that it was inconceivable that your god could possibly be inferior to others' gods. This kind of radicalism could only lead to reckless confrontation. And it did.[3]

Enter modern science. Let's tag 1609 as the year of the great launch. In Italy, Galileo Galilei built his telescope and proceeded to scan the skies as no one had done before. In 1610, he would publish *Sidereus Nuncius* (*Starry Messenger*), telling the world that the cosmos is very, very different from what everyone had been imagining, for, well, essentially forever. With his new tool, Galileo collected compelling evidence that the Sun—not the Earth—is the center of the cosmos, just as Copernicus had suggested half a century before. After three thousand years of Earth-centered astronomy, we were kicked out of center stage, becoming the inhabitants of a mere celestial wanderer, like Mars and Venus. People were perplexed. Could the Sun, the giver of light and life, adored in pagan rites condemned by the Church, truly be the center of it all?

Times were changing. New ideas were bubbling up much too fast for most people to digest. Before this shake-up of the cosmic order, things had made sense: Earth was static in the center; people, rocks, clouds, everything was made out of the four essences: earth, water, air, and fire. The Moon's circular orbit marked the boundary between the earthly and the ethereal. Celestial bodies, including the Moon, were made of a fifth essence, the aether, perfect and unchangeable. The cosmos was like an onion: the planets, the Sun, and the stars all orbited the central Earth in concentric crystal shells. At the outermost orbit was the empyrean sphere, the realm of God and His angels. Life's goal was to be pious and pure so as to ascend to the eternal grace of heaven after death. The vertical ordering of the cosmos was mirrored in people's lives and aspirations.

That same year, 1609, Johannes Kepler published *Astronomia Nova* (*New Astronomy*), where he demonstrated, based on the very precise data gathered by the Danish astronomer Tycho Brahe, that Mars moves around the Sun in an elliptical orbit. A few years later, he argued that the same is true for all planets, including Earth. The circle, this most perfect of shapes, venerated since antiquity for its beautiful symmetry, was no longer the standard celestial orbit. The skies were imperfect.

When Newton's *Mathematical Principles of Natural Philosophy* was published in 1687, the new scientific paradigm was sealed. The *Principia*, as the book is known, was the dawn of a new way to think about reality and our place in it, one that left only so much room for religion. Even though Kepler, Galileo, and Newton, each in his own way, were deeply religious men, their legacy would be a world that needed less and less of God's interference. The more science explained Nature, the less God was necessary. Many felt usurped of their faith, seeing their beloved, all-powerful God squeezed within ever-tightening gaps. There was, however, one gap that resisted any squeezing: the mystery of Creation. Even Deists such as Voltaire, Benjamin Franklin, and Thomas Jefferson, who repudiated any direct divine interference with the world and men, admitted that God was responsible for Creation. Theirs was a watchmaker God, the creator of the cosmos and of the laws that rule the behavior of material objects. The goal of science was to unveil these laws, to decipher Nature's hidden code.

During the early 1800s, the tail end of the Enlightenment, the great French mathematician Pierre Simon de Laplace presented a copy of his masterful *Celestial Mechanics* to Napoleon. In it, Laplace offered a description of the various motions of the solar system, including the details of planetary orbits and their stability. He even proposed a model of how the gravitationally bound system of the Sun and the planets could have emerged from the collapse of a giant cloud of matter. In contrast, the formation of the solar system was something that Newton, approximately one hundred years before, couldn't have conceived of. To him, God was responsible for placing the planets into their orbits and for pushing them into motion. Laplace's work was the embodiment of the clockwork universe, wherein all facets of existence could be reduced to a set of precise mathematical equations. He is famous for saying that a supermind capable of knowing the positions and velocities of all atoms in the cosmos in one instant could predict the future of all that existed, including that of people and their lives. The supermind could predict that I'd write this book and that you'd read it. Not much room left for improvisation here. Faced with such a grim scenario, one can hardly blame the Romantics for rebelling against this abuse of reason. Where would love and spirituality fit in such a machinelike, deterministic cosmos? Would life be meaningful without the drama of choice, without the possibility of making mistakes?

Napoleon allegedly asked Laplace, after saluting him on his magnificent intellectual achievement, why there was no mention of God in the book. "Sir, I have no need for that hypothesis," responded Laplace. What a blow! The all-powerful Judeo-Christian God was relegated to a mere "hypothesis." I wonder how Napoleon reacted to that. He probably knew that Laplace was bluffing; the French mathematician was surely aware that even if he could explain the formation of the solar system out of a giant contracting spinning cloud of matter, he couldn't say where the cloud came from or what had prompted its collapse. Laplace, like so many before and after him, grew quite cavalier with the limitations of his theory.

This triumph of determinism didn't last long. As time marched on, it became increasingly clear that science was not able to explain each of the myriad details at work in Nature. Not everyone was convinced, even as evidence mounted. Some still aren't. In fact, by the late nineteenth century, many influential physicists claimed that their work was nearly done, that all that was left was to fill in some unimportant details. They had conquered the laws of mechanics and of gravity; they had developed the theory of electromagnetism describing how electric charges and magnets interact and how this knowledge can be used to trigger a new, post-steam-engine industrial revolution based on electric currents, circuits, batteries, and electromagnetic motors. The radio, the lightbulb, the telephone, and the telegraph were invented; science was reshaping society at a dizzying pace.

Yet problems started to emerge that took the fizz out of the champagne. The aether was not detected. No one understood why a heated object glowed with a color determined by its temperature, as we see with electric stove burners (red at about 1,000 degrees Celsius) or the Sun (white-yellow due to its 5,500 degrees Celsius temperature at the surface). Another mystery concerned electrically charged metals: when illuminated with ultraviolet light, they would lose their charge; when illuminated with yellow, red, or light of other colors, nothing happened. Application of the tools at hand—Newtonian mechanics and electromagnetism—led to abject failure.

In 1905, Einstein daringly proposed that light was more than simply a wave that could propagate in empty space; it could also be described as being composed of particles, later called photons. This somewhat crazy idea—which Einstein himself thought his most revolutionary—explained the mystery of why ultraviolet light can neutralize charged metals: having higher energy than visible light, its photons could punch the extra electrons out of the metal plate. Many physicists refused to accept such contrarian views. Wave and particle at once? How could that be? Yet Einstein's model explained the data. A quiet panic set in. Science appeared to be losing its grip on reality. Or else, reality was much stranger than everyone had anticipated.

Something was missing: a new description of matter and its properties. It took the first three decades of the twentieth century to sort things out, but eventually this new description of matter—quantum mechanics—was developed. Previous mysteries, and many new ones, were beautifully explained. But the price was high. Precious notions had to be tossed out. The world of the very small, it became clear, is nothing like ours. Strange things can and do happen there all the time. Chief among them is that particles never sit still; they jitter incessantly, as if some inherent discontent animates them. This jittering, encapsulated in the famous Heisenberg Uncertainty Principle, leads to many remarkable consequences: whereas we can measure simultaneously the velocity and position of large systems such as balls and cars with arbitrary precision, the same is not true of atoms and electrons. Their manic jittering bungles things in insurmountable ways. And it's not because our measuring devices are not good enough. The uncertainty is here to stay, a trademark of the quantum world. In the atomic realm everything fluctuates. For example, we can measure the position of an electron under the exact same conditions a million times and each time the result will be different. What we must do, instead, is to average the values of many measurements of the electron's position. Mathematically, this translates into probabilities of finding the electron here or there. The equations of quantum mechanics give us probabilities and not certainties. Laplace's supermind and all its implied strict determinism is gone with the quantum. But here I must add that probabilities don't mean that quantum mechanics is flaky. Quite the opposite: when applied to atoms and particles it gives us highly accurate descriptions. In fact, all of our digital appliances, without which it would seem impossible to survive nowadays, owe their functioning to our ability to predict and control the behavior of electrons, light, and other quantum systems. What started as an annoyance launched a deep revision of our worldview. Reality *is* stranger than everyone had anticipated.

17 LEAP OF FAITH

The quantum revolution had a huge impact on our understanding of the Universe. Apart from using atomic physics to predict the existence of the background microwave radiation detected by Penzias and Wilson, Gamow and his successors used the new knowledge of nuclear physics to construct a scenario where the nuclei of the lighter elements would have been forged during the first three minutes after the bang. Hence the title of Weinberg's inspiring book, *The First Three Minutes*. Originally, Gamow wanted *all* chemical elements to have been forged in the first few minutes of cosmic existence. However, some of his assumptions were not quite right. It took more than a decade and the work of Fred Hoyle and others to clarify the issue. Although lighter nuclei were indeed forged during the cosmic infancy, heavier nuclei are forged in exploding stars. The stuff of life—carbon, oxygen, nitrogen, etc.—comes from the furious nuclear fusion that happens as a star succumbs to its own gravitational embrace.[4]

We should pause to reflect upon the statement above. I find it absolutely remarkable that applying nuclear physics and gravity together, we can explain the origin of the chemical elements. We understand why the lightest elements must be made in the throes of the hot early Universe and the heavier ones in exploding stars. The calculations are precise. They predict the relative abundances of the elements, that is, how much more hydrogen we should find in the cosmos than, say, lithium—both made in the first three minutes, or why iron is more common than uranium—both made in stars.* In rounded numbers,

* Every element is characterized by the number of protons in its nucleus. Neutrons, also part of the nucleus, play an important role stabilizing the interactions responsible for keeping the nucleus together. The arrangement of protons and neutrons of each chemical element has an associated binding energy, defined as

hydrogen makes up 75 percent of matter, while helium, the second-lightest element, makes up 24 percent. The rest of the chemical elements, familiar from the periodic table, from lithium to carbon to uranium, make up only 1 percent. The stuff of life is in the minority. It is a testimony to the solid foundation of the Big Bang model that these predictions are spectacularly confirmed by observations. It gives us confidence that we understand, at least in broad brushstrokes, the history of the Universe down to its first minutes.

Of course, we want to go further toward the beginning. As we retreat in time, the Universe gets hotter and denser. Material structures break down to their fundamental constituents. The chain of events is clear: molecules break down to atoms, atoms to free electrons and nuclei (at 400,000 years), nuclei to free protons and neutrons (at approximately one minute). Note the enormous jump in time between the dissociation of atoms and of nuclei. This is due to the huge difference between the force that binds atoms and that which binds nuclei together. In atoms, electrons are electrically attracted to protons. In nuclei, protons and neutrons are attracted to each other via the strong nuclear force, which is about one hundred times stronger than the electromagnetic force. This explains why protons stick to other protons in nuclei even though they have the same electric charge and thus repel each other: the glue of the strong force overwhelms the electric repulsion between protons. The difference in strength between the two forces also explains why nuclei with more than 100 protons are highly unstable. Specifically, things get bad after uranium, with 92 protons, although plutonium, with 94, still occurs naturally.[5]

Less than a minute backward, another big change happens: the heat is so intense that not even protons and neutrons can bind into nuclei. We must use particle physics to describe the way matter interacted. It is here that one of the key concepts of modern physics enters the game: the breaking of the symmetries used to describe how the fundamental particles of matter interact. The topic is so important that it deserves its own part, the next one. For now, I want

---

the amount of energy needed to break the nucleus apart. Of all chemical elements, the nucleus of iron is the most tightly bound. As stars forge different elements during their final moments, this property of iron gives it an enormous advantage, making it one of the most abundant among the heavy chemical elements.

to take another jump backward in time, a very daring one, all the way to the beginning. How close can we get?

For the moment, we can dispense with the details of particle physics. It is amazing how little is needed to describe in general terms the earliest stages of the cosmos. The key observation is that, following the premise of the Big Bang, the closer to the beginning, the smaller the Universe. Hubble, the man who revealed to the world that galaxies are moving away from each other, used his discovery to predict the age of the Universe, the time since the Bang: what is needed is a measure of the galaxies' recession velocities. Playing the movie backward, we get to a time when they were all on top of each other. That's the beginning. Unfortunately, measuring the distance and velocities of runaway galaxies is very hard. Hubble got a bad result, about two billion years, less than the then-known age of the Earth. How could the daughter be older than the mother? It couldn't. Better measurements during the following decades settled the issue. We now know that the oldest stars fit well within our 14-billion-year-old cosmos.

To move closer to the Bang we need to take an enormous leap of faith: we must assume that the physics describing the much tamer conditions accessible to measurements today can be extrapolated to the violent mess close to the beginning. The most advanced particle accelerators manage to collide particles at energies comparable to those about one-trillionth of a second from the Bang, that is, 0.000000000001 second from the beginning. This time may seem ridiculously small, and, indeed, for our standards it is. But a photon, a particle of light, can cross a proton about one trillion times in this small time interval, a relative eternity.[6] Even if one-trillionth of a second is quite an early time, to probe the cosmic origin we must go much earlier, beyond known physics. The assumption—the leap of faith—that we can extend in this direction is fair as long as our speculations include predictions and tests that can, at least in principle, be realized in the near or even not so near future. In the face of negative results we should be prepared to let go of even our dearest notions, as Nature doesn't care about our predilections. We should also be wary of speculations that go too far into the unknown, losing any link with testability. As I wrote earlier, physical theories that cannot be tested—or that can be conveniently rescaled to be always outside the realm of testability—should not be part of the scientific canon.

18 THE JITTERBUG COSMOS

Back to the beginning, we come to a time when the Universe itself has dimensions comparable to those of an atom. This is the quantum cosmology era, the point where our theories break down. We hit a very serious conceptual barrier here, which many physicists have been trying to overcome for the past four decades. Fascinated with this question since my graduate student years, I have struggled with it myself. The problem is Einstein's theory of general relativity, which he completed in 1916, eleven years after the special theory that did away with the aether. The general theory provides a new explanation for gravity as the curvature of space around mass concentrations: the bigger the mass and the smaller the volume, the more space is bent around it, and the stronger the gravitational pull toward it. Picture a gymnast jumping on a trampoline. As she jumps up and down, she stretches and contracts its elastic surface in different directions. Einstein's idea was that matter can bend space, and thus deform its geometry away from flatness.

Time is also affected. Stronger gravity slows time down. A clock at the surface of the Sun, if it could work there, would tick more slowly than it does on your wall. The interconnectedness of space and time is one of the most striking consequences of both theories of relativity, special and general. What we commonly perceive as two different entities are, in fact, part of a single structure called space-time. According to the theory, it's best to picture time as a dimension, as we do space. We move east-west, south-north, and up-down, along the three dimensions of ordinary space. For the time dimension, it's past and future. Physicists talk of distances in space-time the same way people do in normal space. It's useful to think of space-time as a peculiar kind of elastic sheet linking space and time, somewhat like the gymnast's trampoline. Space can shrink, time can slow down, and vice versa. The net effect depends on the source of gravity (for the general theory) and on how two or more observers measuring distances and time intervals are moving relative to each other (for the special theory).

In practice, our speeds are too slow compared to the speed of light, and Earth's gravity doesn't bend space-time much. The changes resulting from both the special and the general theories are all but negligible. From our limited three-dimensional perspective, we have a myopic view of reality and split space and time into two different entities. But if we could wear glasses with the prescription that the theory of relativity dictates, we would see the union of space and time in all its glory. In the absence of such glasses, we have math. The strange stretching and shrinking of space and time is well understood and experimentally confirmed. Recall that space and time are descriptive tools we create to quantify the transformations of the natural world. Having them behave in more plastic ways is just the small adaptive change needed to improve our description.

As we move toward the beginning, the shrinking of space forces us to consider how quantum physics will affect the young Universe: loosely speaking, the cosmos had atomic dimensions. Herein lies the challenge; so far, we have not been able to construct a theory of gravity compatible with quantum mechanics. We know that, early in the cosmic history, the jittery nature of the atoms must be brought to the cosmos itself. But how? Since in the world of the very small everything fluctuates, at short distances space and time will also fluctuate. Measurements of distances and of time intervals, which we take for granted, will become probabilities. Picturing space-time as a rubber sheet again, at the quantum level we would see oscillations that twist and contort it in myriad ways. Time runs amok. The consequences are mind-boggling. Without reliable measures of distance and time, or of how to interpret them probabilistically, the whole edifice of physics crumbles. The notion of a phenomenon as something that happens in space and time becomes meaningless.

Just as with the situation at the close of the nineteenth century, a new idea is needed, a new theory of space-time that marries quantum theory to Einstein's gravity. This new formulation, whatever its structure, *must* reproduce the known observations at later cosmological times: cosmic expansion, abundances of the light nuclei, homogeneity of the cosmic microwave background. This is its bare minimum requirement, consistency with the Universe we live in. Modern candidates include superstring theories and loop quantum gravity, although neither can convincingly claim to reproduce the known Universe. Superstrings have the distinction of also allegedly being a Theory of Everything, the modern incarnation of the Final Truth. They are based on a radical change of the prevalent atomistic perspective: the fundamental building blocks of matter are not individual pointlike particles like electrons but tiny wriggling strings made of nothing else. Different particles of Nature, such as electrons and photons, result from different vibrating patterns of these fundamental strings. Proponents see it as the culmination of reductionism, a fully unified theory of Nature, although no one knows exactly how that will be accomplished, in spite of three decades of hard work by some of the brightest minds on the planet. I see it as the modern incarnation of the Pythagorean myth, the search for a monotheistic explanation of Nature based on geometric arguments: symmetry changed from a useful tool to overarching dogma. But this is not how I saw things as a graduate student in England, when I was immersed in the search for the Final Truth. Nothing could be more fascinating, no theory more compelling. All six papers I wrote during my graduate studies, plus quite a few others in the years following my Ph.D., dealt with different aspects of superstring unification. I will explain why I changed my mind in Part III. First we need to explore how quantum theories of gravity—proposals to relate quantum mechanics and general relativity—describe the properties of a quantum space-time.

If we assume that quantum mechanics must survive at early cosmological times, and we know of no reason why it shouldn't, quantum jittering will cause fluctuations in space-time itself. We go back to the creation narrative that opens this book, and consider a bubbling soup of geometries, possibly of all types, coexisting in the endless multiverse. If the cosmos were a musical orchestra, all symphonies, all possible sounds would be playing at once: organized and chaotic, from the most sublime to the most absurd, from long elaborate musical phrases to short cacophonies. All, of course, without a conductor. Some versions of superstring theories predict a near-infinite sea of cosmic possibilities, the "landscape." The name is very suggestive: imagine a vista with valleys and peaks stretching in the distance. Different valleys in the landscape correspond to different universes, each with its own properties, possibly even with distinct values for the constants of Nature. In some the speed of light may be greater than in ours; in others, lesser. In one, light doesn't exist; and in another, electrons have different electric charges and masses. There is no time in the landscape; it represents only a mapping of potential universes, each a possible solution to the equations of superstring theory.[7] Contrary to the expectation of many string theorists, the picture that emerges from the landscape is far from "elegant." If the idea persists and no valley can be shown to be more compelling than all the others, the dream of uniqueness, of finding *the* solution that explains our Universe, has to be abandoned. In an ironic twist, the very theory that was supposed to bring Oneness into physics, the theory that was supposed to reveal the Final Truth to the world, ends up showing how pointless the notion is.

Back to the beginning, even if we do away with a string-inspired landscape, the basic notion that at the quantum level space-time can be described as a soup of fluctuating geometries should survive in one form or another. All competing versions of quantum gravity theories retain the jittering of space-time.[8]

Accordingly, our cosmos would have sprung from this quantum soup, a bubble with a highly unlikely combination of natural constants to generate ever more complex structures, including living ones that can reflect upon their existence. Back to the orchestra metaphor, it is hard to resist the temptation of calling our music beautiful. And thus rare.

Something is missing from our discussion. Space and time, albeit fascinating, are only part of the story. A universe without any matter is quite uninteresting. But what kinds of matter can we put into these fluctuating cosmic bubbles? In the absence of data, choices proliferate. We simply have no clue. No fossils of the earliest times have been found. There are, of course, candidates, but no concrete observations. This is when one may feel that it's time to throw in the towel and call it quits, perhaps go work on laser physics or fluid dynamics, areas where leaps of faith are not so dramatic. However, not all is lost. Our Universe, if we use the right tools to "listen," tells us many things about its past. The strategy of modern cosmology has been to measure the properties of the present Universe to infer clues about its infancy. The Big Bang model is then adjusted to make it compatible with the combined observations. It is a work in progress.

19 THE UNIVERSE THAT WE SEE

The first key observation about the Universe is also the most obvious. It's really big. How big? We have to be very careful here. All we can talk about is the part of the Universe that we can see, that is, observe with our different telescopes and antennae. Recall that the speed of light limits information exchange: since nothing can travel faster than light, we can only receive signals from regions within our causal past, that is, information that has had time to travel to us at the speed of light since the beginning of time. For example, the Sun is about eight light-minutes from Earth; if it exploded right now, it would take eight minutes for us to find out, our last eight minutes, by the way. The nearest star, Alpha Centauri, is about 4.37 light-years away: when we see it in the sky, we are actually seeing it as it was 4.37 years ago. This is the time it takes for light from its surface to reach us.* Zooming out of the Milky Way we meet Andromeda, our neighboring galaxy at about two and a half million light-years away. The light we see began to travel toward us when our hominid ancestors were spreading across the African savanna. To look into the depths of space is to look back in time.

How far back can we go? Very. Modern-day telescopes probe electromagnetic radiation in all wavelengths, from the very short gamma rays to the very long radio waves. The farthest (and oldest) sources are very faint or invisible. Astronomers "see" them with extremely powerful telescopes, often combining several observa-

* A light-year is the distance covered by light traveling in empty space for a year. It comes to about 9.5 trillion kilometers, or 5.8 trillion miles, or 63,000 times the average distance from Sun to Earth.

tions into a single search that mixes different kinds of light, from visible and ultraviolet to radio. For example, early in 2008, astronomers at Rutgers and Penn State universities in the United States announced the discovery of baby versions of spiral galaxies like our Milky Way at a distance of 12 billion light-years from Earth. Considering that the Universe itself is just under 14 billion years old (13.73 billion years old, to be precise), light from these galaxies traveled for 12 billion years to reach us. It left almost 8 billions years *before* the Sun and Earth even existed.

Can we see all the way to the beginning? Is it just a matter of building more powerful telescopes? Unfortunately, no. Seeing a distant object means capturing photons coming from it. In turn, this means that light must travel relatively unimpeded from the object to our telescope or radio antenna. As we move back in time, we hit an opaque wall, the time when the first atoms were made. Before this time, photons were actively engaged in breaking the electric binding between electrons and protons and were not free to travel across space. We have seen that this is when the cosmic microwave background radiation was released, four hundred thousand years after the Bang. To probe the Universe before this time, we must search for clues other than electromagnetic radiation. We have encountered examples already, the light nuclei of helium and lithium synthesized when the cosmos was about a minute old.

In spite of this opaque wall into the very young cosmos, we can combine available astronomical observations of galaxies and of the cosmic microwave background to make several key statements about the Universe. First and foremost, as long as you look at large enough regions, the Universe seems on average to be the same everywhere. True, looking at a starry night doesn't seem to corroborate this statement. In our urge to seek order, we fill the sky with our fantasies and longings, finding lions, crabs, cooking pots, heroes, and dragons, instead of equidistant stars. Those patterns emerge from projecting the stars onto the two-dimensional celestial dome (which is not really there, either). In reality, distances of dozens, hundreds, even thousands of light-years separate stars in constellations, which are distributed without any obvious pattern to them. When we say that on average the Universe looks the same we mean that when we look at the objects distributed within distances of hundreds of millions of light-years, things look similar. Imagine a crowded beach on a hot Sunday morning, say Ipanema in Rio or South Beach in Florida. From far away, we see a multitude of heads, making up a fairly uniform mass. Only as we zoom in can we start to distinguish the varying details of the people themselves: some blondes and redheads, some teenagers sprawled on towels furiously text-messaging, a child building a sand castle, another collecting shells.

That the Universe at large is homogeneous and isotropic (the same everywhere and in all directions) simplifies things tremendously. If we are only interested in the overall cosmic history, we can dispense with the details of what happens in this or that galaxy and focus on the Universe as a whole, that is, on how it evolves in time. This is the essence of the Big Bang model, to focus on the overall time evolution of the cosmos. Local details, such as the formation of galaxies and the birth of stars, are a separate issue.

Can we trust the approximation that the Universe is homogeneous? Fortunately, yes. Both astronomical data and the cosmic microwave radiation back it up. The microwave background, in fact, is incredibly homogeneous. Its temperature shows only the tiniest variations about its mean of 2.73 degrees above absolute zero, defined as the coldest possible temperature. Had matter been distributed in very large clumps when the Universe was four hundred thousand years old, the curvature of space about them would have affected the photons of the microwave radiation, causing them to lose and gain energy, like children going up and down slides. These gains and losses would translate into large fluctuations in the microwave temperature, contradicting observations.

In 1989, a NASA satellite called COBE (Cosmic Background Explorer) went up to measure the properties of the microwave radiation to a precision much higher than that achieved by Penzias and Wilson in 1965. The results were astonishing. Not only did the temperature display an amazing degree of homogeneity, but the fluctuations about its mean value are extremely small, only about one part in one hundred thousand! Compare this to the surface of the Earth: if we shrink all mountains by a factor of one hundred thousand, Mount Everest (the largest "fluctuation" about Earth's average surface height) would measure less than a meter. (More precisely, 0.088 meter, or 0.29 foot). An average person would be about forty times smaller than an amoeba. The Earth would look very flat! Now, imagine aliens observing this flat Earth with their instruments and still being able to discern mountains, valleys, and skyscrapers. To achieve this kind of precision in a space mission is a tremendous technological feat. In 2006, another pair of astrophysicists, John Mather and George Smoot, received the Nobel Prize for their leadership in the COBE project.

Another very important property of our cosmos surfaced through observations of galaxies and the microwave background. Apart from having an amazingly homogeneous distribution of matter and radiation, the Universe is also flat. Sounds like quite a boring place, it's true, although if it weren't for these two properties we wouldn't be here. What does it mean, "the Universe is flat"? *What,* exactly, is flat? We go back to Einstein's general relativity and to the relationship between geometry and matter. There are only three homogeneous and isotropic types of geometry: a flat one, familiar from two-dimensional tabletops; closed geometries, such as the two-dimensional surface of a ball, always curving in the same way; and open ones, like two-dimensional horse saddles, that curve in two opposite directions, downward for the legs and upward in the back and front. These examples refer to two-dimensional shapes for a reason. We have trouble seeing three dimensional surfaces. It's easy to visualize a curving ball or a saddle because we can see the surfaces from a distance, that is, from a third dimension. That's what we need to move about in the world. To see a three-dimensional surface we would have to step out into a fourth spatial dimension. Only mathematics can do that.

Geometry is about measuring distances. On a flat tabletop, you can measure the distance between two points by linking them with a string. You can do the same on the surface of a ball or on a saddle. Soon enough, you will find out that geometry has different properties on these surfaces. For example, the famous law of Euclidean geometry, that the sum of the internal angles of a triangle is always 180 degrees, is only true in a flat geometry. In a closed geometry the sum is larger, while in an open one it's smaller. Measuring distances and angles between points provides a way to distinguish between geometries. A flat universe is meant to obey the rules of flat Euclidean geometry.

In practice, though, the cosmic distances are so huge that astronomers use a different method to measure it. Since the Universe is homogeneous at such large scales, we can study how it evolves as a whole in time, without paying attention to local deviations. This is what Hubble started to investigate in 1929 and that nowadays has reached enormous precision, thanks to the Hubble Space Telescope and its Earth-bound cousins. According to the Big Bang model, there is a deep relationship between how a universe evolves in time and the amount of matter it contains. In other words, the rate at which the Universe expands (or contracts) tells us about how much matter it has. A universe with lots of matter will exert a strong gravitational pull upon itself and have trouble expanding. A universe with little matter will be able to expand more freely. The same kind of reasoning applies to launching rockets from the Earth and the Moon. Since Earth's gravitational pull is larger (by about six times) than that of the Moon, it takes more energy to launch a rocket from here. This is why you see videos of astronauts hopping lightly about the Moon. Measurements of the speed at which galaxies move away from each other and of the properties of the microwave background decisively point toward a flat Universe, with current precision of about 98 percent, slightly skewed toward a closed geometry. We can't state for sure yet, but better measurements will probably bring the result even closer to flatness.

The flatness of the Universe means that the total amount of matter and energy is delicately balanced between being too much and too little, like a needle standing upon its end. A bit more of it, and the Universe would be closed. A bit less, and it would be open. The resulting difference in behavior is dramatic. A closed universe will tend to collapse upon itself after expanding for a while. The more matter, the sooner the "big crunch." In contrast, both flat and open universes will keep expanding, although an open universe expands faster than a flat one. In more precise terms, a flat universe has a very specific density of matter and energy, called the *critical energy density.* For our Universe, it is only about ten hydrogen atoms in a cubic meter. On average, our cosmos is very empty.

We can state with confidence that the Universe we can see is homogeneous and flat. And there should be more of it beyond what we observe, beyond what we call our horizon. The name is suggestive of what happens at sea, where the horizon marks the line at which sky and water meet in the distance rather than the end of the ocean. The same is true of the cosmos. Beyond our horizon there is more universe, forever beyond the reach of our telescopes. If we picture the visible portion of our Universe as an enormous bubble centered in the Earth, the horizon is the farthest point in our past. This, in practice, is the size of our observable Universe. Putting the numbers in, we get a radius a little over three times the age of the Universe in light-years, about 46 billion light-years.[9]

20 THE FALTERING BIG BANG MODEL

Now that we know what the Universe at large looks like, the challenge is to understand why. This is where the Big Bang model starts to falter. The cosmic homogeneity and flatness are not predictions of the model, only input parameters used to pin it down. If you are going to build a house, you need bricks and mortar. You also need the blueprint to put the bricks and mortar together into a house. Different blueprints use the same brick and mortar but lead to different houses. In cosmology, the bricks and mortar are the laws of physics. Different blueprints using the same laws of physics lead to different universes. The question we have to answer is, Why *this* house? What determines the cosmic blueprint of *our* Universe, the fact that it is homogeneous and flat?

The problem is more pressing than it seems. We stated that the cosmic microwave background was remarkably homogeneous, at a temperature of 2.73 above absolute zero.* This creates a bizarre paradox. Picture yourself at the center of our visible Universe, equidistant from the horizon. You have a powerful microwave antenna and can measure the temperature of the radiation in all directions in the sky. Let's say you measure the temperature at two diametrically opposite directions, such as east and west. You will get the same result, 2.725 Kelvin. Is this what should be expected? In general, equal temperatures in different places means that the matter in these places has been in contact. For example, if a bucket of boiling hot water is poured into a bathtub half filled with cold water, it will

*The best value at present is a temperature of 2.725 Kelvin. We can round it up to 2.73 Kelvin for economy.

take some time for the water temperature to equalize again. This new temperature, called the *equilibrium temperature,* will be somewhere between the cold initial temperature of the water in the bathtub and the boiling temperature of the water in the bucket. The time it takes for the water to reach the new equilibrium temperature is called *equilibration time.* It depends on collisions between hot and cold water molecules: hot molecules move faster and kick harder, making cold molecules speed up (while they lose energy and slow down). Temperature is simply a measure of the molecular speeds. After many collisions, the speeds will all become similar. Thermal equilibrium is a situation where, on average, molecules have the same speed and thus the same temperature. There will, of course, be fluctuations about the mean, but they are not very important.

The whole of space within our cosmic horizon is like a spherical bathtub filled with radiation in thermal equilibrium. The problem is that while the points in the far east and the far west are equidistant from us, they are *twice* as distant from each other and hence outside each other's horizon. Remember, we are measuring distances from our location in the middle of the horizon volume. Unlike the hot and cold molecules in the bathtub, if the points are that far away, they could never have been in causal contact. Yet their temperature is the same to the amazing accuracy of one part in one hundred thousand. As no particles of matter or radiation can travel faster than light, this equilibration mechanism was certainly unusual. Does it contradict the laws of physics? This mystery is called the "horizon problem." It doubles with the "flatness problem," the absence of a mechanism justifying the cosmic flatness, as the two most serious limitations of the Big Bang model. We need a new idea, an idea that transcends common sense.

21 BACK TO THE BEGINNING

The Universe that we see is homogeneous and flat. The smoothness of the microwave background temperature tells us that this has been true from at least four hundred thousand years after the Bang until today, that is, for most of the cosmic history. In fact, the Universe must have been this way when it was a few minutes old, during the fusion of protons and neutrons into the first light nuclei, the period known as *primordial nucleosynthesis*. Only a homogeneous and flat Universe would have generated light nuclei with abundances that agree with our observations. This means that we have a good understanding of what was happening in the cosmos when it was only about a second old, a remarkable achievement of modern cosmology.

This is where Weinberg's *The First Three Minutes* ended. There were a few remarks about earlier times, but Weinberg admitted they were vague. The sense, though, was that there was much new territory to be explored within the first second of cosmic history.

Having established the overall cosmic properties, we must now go beyond the Big Bang model and tackle a harder question: Why is the Universe this way?

In 1981, the year I graduated in physics from the Catholic University of Rio, Alan Guth, now a professor at the Massachusetts Institute of Technology, came up with a radical idea. What if the young Universe expanded incredibly fast for a short while, faster than the speed of light? Then it's possible to show that . . . "Wait a second!" the attentive reader protests. "I thought nothing could move faster than light." Matter and radiation can't, it's true. But space can. There is nothing in the laws of physics that prohibits it. If we imagine again lampposts along a straight road with each lamppost representing a galaxy, the expansion of the Universe is equivalent to the road itself stretching as if it were made out of rubber. As it does, it will carry the lampposts along with it at its stretching speed, whatever it is. The light coming out of the bulbs, however, will still move at its normal constant speed. The speed of light is the ultimate speed with which information can travel. Space doesn't share this kind of limitation.

Guth's idea was that space itself stretched incredibly fast right at the beginning of time. Two points initially close together would have been pushed away faster than the speed of light. In order to work, the expansion should have been exponentially fast. Hence the name, *cosmological inflation.*[10]

Two obvious questions come to mind: How does a faster-than-light cosmic expansion solve the problems of the Big Bang model? What could have caused space to expand at such speed?

First we examine how inflation helps the Big Bang, since that's simpler. What happens when we blow up a party balloon? If we focus on a small patch on its two-dimensional elastic surface, we see that it flattens as the balloon grows. Now imagine an exponential expansion by a factor of sixty, which is the number cosmic inflation needs to be efficient. The balloon's surface becomes fantastically flat. For comparison, if Earth expanded by the same factor, a protuberance such as Mount Everest would appear to shrink to about one-millionth of the size of a proton. Likewise, after such a dramatic expansion, the observed Universe would have become very flat. So much for the flatness problem.

An exponentially fast expansion also solves the horizon problem. Consider a region small enough that matter and radiation within it are in thermal contact. This means that they can interact through fast collisions to establish thermal equilibrium, determined by a (near) uniform temperature. As the region is stretched exponentially fast to a fantastic size, particles of matter and radiation in its interior remain locked in equilibrium. Inflation makes it possible for the whole of our observable Universe to have emerged from a tiny region at the beginning of time, a region small enough that all matter and radiation within it were in thermal equilibrium. In a fraction of a second this region was stretched exponentially by at least a factor of sixty so as to encompass the Universe we see. The faster-than-light expansion of space overcomes the limitations of causality without violating any physical law.

Inflation is a beautifully simple, efficient idea. We can reverse the argument and state that the flatness and homogeneity of the observed Universe are *predictions* from inflationary cosmology. The inflation theory makes other predictions that, as we will see, are also in agreement with observations.

Simple theories with great explanatory power are enormously attractive to scientists. The aesthetics of theories appeared in philosophy at least as early as the tenth century. William of Ockham, a leading scholastic philosopher and Franciscan friar who lived in fourteenth-century England, was the first to develop a method to organize and simplify theories that seek to explain various aspects of reality. "Plurality ought never to be posited without necessity," he wrote. When deciding among competing theories, scientists apply a distinguishing criterion known as Ockham's razor: given two valid explanations for the same set of phenomena, the simpler one is assumed to be true. Famous examples include Newton's theory of universal gravity and Darwin's theory of evolution. Other explanations, even if possible, are not as economical and hence probably incorrect. Given that, even theories that are deemed correct work only within their limit of validity. Pushed too far, they will fail.[11]

Useful as it is, Ockham's razor cannot act alone. Irrespective of how aesthetically attractive a theory is, the final deciding factor is always data. A simpler theory, initially considered correct, may be unable to explain new observations. In such cases, scientists either develop new formulations or revive older ideas, which may previously have failed Ockham's test. When searching for new theories, it is important to keep in mind that Ockham's razor is only a selection tool and should *not* be used to decide whether or not a given theory is correct. Nature has the final say: simpler is not always better. In the heat of invention it is easy to let aesthetic values function as choice criteria, confusing a "beautiful" or "elegant" idea with a correct one. Contrary to our aesthetic longings, beauty is not always truth.

22 EXOTIC PRIMORDIAL MATTER

In spite of its compelling explanatory power, inflationary theory is not quite there yet, as it still can't be validated beyond doubt. The major stumbling block is not what the theory can explain but how it works, that is, the physical mechanism behind the fast expansion. This is where things get murkier.

What could cause a faster-than-light expansion in the early Universe? To answer this question, we must recall that according to Einstein's theory of general relativity, matter and energy tell space how to bend. Different kinds of matter will have different effects on the geometry of space. In practice, even with cutting-edge computers, solving the relevant equations can be a very cumbersome exercise, as complicated distributions of matter will generate very complicated geometries.

Fortunately, things are much simpler in cosmology. As we are only interested in the behavior of the Universe as a whole, all we need is the average properties of the matter and radiation that fill it. In practice, we approximate all kinds of matter and radiation to a gas, such as air. The advantage is that we need only two properties to describe a gas, its *pressure* and its *energy density*. We know intuitively what pressure is: the net force exerted over a surface. A party balloon grows as you blow into it because air molecules bang against the inside of the rubber, making it stretch. This familiar kind of pressure, always positive, is a consequence of the motion of the molecules: the faster they move the higher the pressure is. Energy density means the amount of energy in a volume. Here, since this is a relativistic theory, we must be careful to add all possible contributions to the energy: the masses of the particles (from the $E=mc^2$ formula or, written more instructively, $m=E/c^2$, showing that mass has energy in it), the energy of the particle's motion contributes to the pressure (the faster they move, the more energy they have and the more pressure they exert); and, lastly, what is called potential energy, the energy that may be stored in matter due to its interactions. For example, a stretched rubber sheet or a distended spring has energy stored in it, what we call elastic potential energy. If let go, it snaps back to its strain-free equilibrium position. Similarly, an electron and a proton, when near each other, store electric potential energy due to their electric attraction. As with two lovers saying good-bye at a train station, we have to apply a force to pry them apart.

Modeling matter and radiation as a gas, we feed its energy density and pressure to the equations dictating how the universe expands, and study the possible solutions. What we find is that all reasonable kinds of matter, either moving very fast or very slow, or radiation, which always moves at the speed of light, produce expansions that are slower than the speed of light and hence of no use for inflationary cosmology. As when we step on the car brakes, these ordinary kinds of expansions all have *negative* acceleration and slow down in time. In cosmology, the gravitational pull of matter filling the cosmos provides the "braking." These normal types of matter and radiation—electrons, protons, neutrons, neutrinos, photons—are the ones that fueled the cosmic expansion during the production of light nuclei and, later, that of the first atoms.*

In order to generate faster-than-light expansion, admittedly a strange concept in itself, a strange kind of matter is needed, one that generates *negative pressure*. If large enough, negative pressure can cause the Universe to expand with *positive* acceleration.[12]

Normal kinds of matter get diluted as space expands; their energy density and pressure drop as the volume increases. The exotic matter

*Note that lower pressure stretches space faster. This peculiarity can be blamed on the general theory of relativity. Contrary to normal Newtonian physics, which has nothing to say about the effects of matter and energy in the curvature of space, in general relativity, both energy and pressure can affect the cosmic expansion. And pressure acts counterintuitively. You can think of it as a having a "mass," so that higher pressure (higher "mass") means slower expansion. So, to make space expand faster, the pressure has to be small. This notion is taken to the extreme in inflation.

needed for inflation is different: like a rubber sheet that keeps releasing energy as it contracts and then immediately stretches back to its initial state, its energy density and pressure remain approximately constant as space expands.

Here, the exasperated reader may say that physicists are indeed losing their minds and wasting their time (and his) discussing nonsensical notions. Fortunately, in this case at least, this is not true. There are certain kinds of matter that do generate negative pressure. They are bizarre, yes, but not implausible. In fact, the main inspiration comes from a very familiar class of phenomena known collectively as *phase transitions*. We know that liquid water turns to ice when cooled below its freezing point: a transition from a liquid to a solid phase. As the temperature is lowered, water molecules start to aggregate in small clumps that have a certain order to them, tiny ice crystals. As we will see in Part III, this idea can be adapted to the way elementary particles of matter interact: they too can experience a qualitative change in their properties and behavior as the temperature is lowered or raised.[13] Just as the spatial effects of electricity and magnetism can be described in terms of a field, these qualitative changes in the behavior of matter also have an associated field, called a "scalar" field. Scalar fields are the bricks and mortar of modern cosmology. They are used to build not only unified theories of matter and forces but also models of inflation's ultrafast expansion. They are the cornerstone of theories describing the origin of matter—possibly the sources of matter itself—and the origin of galaxies. Most modern research on early Universe cosmology includes scalar fields in one way or another. We must thus spend some time describing their properties and uses.

The unusual name, *scalar field*, comes from mathematics: scalar quantities vary continuously in space but have no direction. Examples are the height of waves on the surface of a lake or the temperature in a room. We can associate a "temperature field" to a room by measuring the temperature at every point inside the room. In contrast, the flow of a river or the blowing wind always have, at each point in space, both a speed value and a direction where they are headed. These are called "vector" fields.

What are these scalar fields? In the so-called Standard Model of particle physics, which combines all our knowledge of the elementary particles of matter and their interactions, they are supposed to represent a kind of matter that interacts with all other kinds of matter. (The Standard Model will be addressed in Part III in greater detail.) These scalar fields (there may be more than one) are a bit like the aether in that they are always there in the background, a medium offering some kind of resistance to the motion of electrons and other particles of matter, as when marbles fall through honey. Since motion is related to inertia—a measure of a particle's mass—scalar fields are thought to affect a particle's mass. In fact, they are supposed to *determine* it. Different particles of matter interact differently—each with its own strength—with scalar fields and hence have different masses: the stronger the interaction the larger the particle's mass. In the Standard Model, the scalar field is called the Higgs, for Scottish physicist Peter Higgs, who proposed its existence and role as overall mass giver.

At the time of writing, the Higgs has yet to be found. There is a very good chance that it, or something like it, will be seen in the Large Hadron Collider (LHC), a huge particle accelerator designed to elucidate the mechanism by which particles of matter get their masses. The LHC is a 27-kilometer circular tunnel buried one hundred meters underground, nested along the Swiss-French border, the largest machine ever built. It will hopefully start operating in earnest in 2010. Even if the Higgs does not exist as an independent particle, the notion that scalar particles and their associated fields play a key role in high-energy physics is here to stay. For example, perhaps two particles interact so strongly that they appear to be a single entity that behaves and *looks* like a particle from a scalar field. Unless studied at energies that match the intensity of their attractive interaction, their collective nature may not be seen. A hydrogen atom is a bit like this. From a distance, it has zero net electric charge, since the charge of the electron cancels that of the proton: the atom looks like a neutral object. Up close at the atomic level, though, the story is quite different. Objects that look simple from a distance may actually be quite complex. Maybe the Higgs is a composite like that, maybe not. Only experiments can tell. In either case, something like it serves inflation well.

23 A SMALL PATCH OF WEIRDNESS

Whether effective descriptions or fundamental entities, scalar fields can produce the positive acceleration needed to fuel the faster-than-light cosmic expansion of inflationary cosmology. If the field is away from its equilibrium state and thus under strain, it will behave like the mythic cornucopia that never stops producing fruit. Until, that is, it reaches its equilibrium, or lowest-energy, state.

In Guth's original model, the all-powerful scalar field that drove the exponentially fast expansion came from a theory that went beyond the Standard Model and attempted to unify the three interactions that determine all the atomic and subatomic properties of matter: electromagnetism and the strong and weak nuclear forces. Such theoretical constructions, which go by the humble name of "Grand Unified Theories" or GUTs, have been proposed for more than three decades as a major step toward the Final Truth. Full unification would be eventually completed through the incorporation of gravity, the fourth and last force. Unfortunately, as we will see in Part III, the main predictions from grand unification haven't yet panned out. To make things worse, Guth's original model didn't work, either: the Universe would never stop inflating. In spite of these setbacks, Guth and others soon realized that the general idea behind inflation doesn't rely on a specific GUT model.[14] A new strategy was developed disconnecting inflation from GUTs. The logic goes as follows: "Forget Guth's original motivation from particle physics; given that we don't really know what was going on during the very first moments of cosmic existence and that we have no experimental confirmation of the validity of GUT models, let's just *assume* that some form of scalar field was present at the time and was able to generate the negative pressure needed to fuel inflation. Inflation is too simple and compelling an idea, it solves too many problems to be abandoned. Whatever drove it, it played the role of an effective scalar field. Details can be worked out afterward, using observational cosmology to constrain possible models of particle physics."

Perhaps it's best to let go of the deep relation between particle physics and cosmology, and use some vague kind of effective scalar field to do the job as opposed to *the* GUT scalar field. This way, if GUT theories are proven false, not an unreasonable possibility in the least, inflationary cosmology still survives. What one gives up, of course, is beauty and elegance, the oneness that so many believe should provide the backbone of early Universe cosmology.

As a beginning Ph.D. student at King's College in London, I was shattered when Guth's original model was shown not to work in 1982, and that in all probability no other GUT scalar field would work, either. If the scalar field that drove inflation is just one field among possibly many others, gone is the uniqueness of the theory, of the connection between unification and cosmology. All boils down to having some scalar field active early on, that is, away from its lowest energy state, like a ball ready to roll down a hill. In current inflationary models, all that is needed to generate our Universe is a tiny region of space filled with a scalar field displaced from equilibrium: just a small patch of weirdness without anything grand or unique about it. The mathematics shows that the negative pressure from a scalar field with the right amount of energy would drive a small region of space into an expanding giant. In the new cosmology, inflation is not necessarily interwoven with the unification of the fundamental forces. The loss of elegance is the gain of generality. Our cosmos does not need perfection to exist. It needs unbalance.

There is still a very important point I haven't touched upon. If the Universe was initially mostly filled with a scalar field, where do electrons, photons, and neutrinos come from? This question, it turns out, is related to another one: how does the Universe decelerate from a superfast inflation to the slower expansion of the Big Bang model? This transition is the subject of much debate today. Since we don't know the details, all we can say at this point is that as inflation progresses the unstable scalar field converts itself and its energy into other particles: in a process similar to radioactive decay, where one particle morphs into two or more, the scalar field becomes other kinds of matter particles. Eventually, these initial decay products convert into more ordinary particles. According to this view, the primordial scalar field that drove inflation would be matter's first common ancestor. This is not as strange as it seems. Particles decay and change into each other all the time. In the subatomic world, instability and transformation are the rule. A lonely neutron, for example, within about ten minutes decays into a proton, an electron, and an antineutrino. According to current understanding, as inflation nears its end, the matter conversion process goes berserk. The scalar field explosively dumps its remaining energy into a maelstrom of particles, filling the cosmos with hot matter. In the modern view, it is this explosive creation of matter at the end of inflation that is associated with the Big Bang: in other words, *the Big Bang is not the beginning.* The details, however, remain nebulous. We don't know which particles the scalar field interacted with, how it decayed into them, and how they eventually became the particles we know of. When inflation lost its connection with GUTs, our explanations for these phenomena evaporated. The best that we can do at present is to develop new models and study their viability.

The history of cosmic inflation, now three decades old, reveals something very important about how physics is done. The original motivation was very clear-cut: there were issues with the Big Bang model, the horizon and flatness problems among the most important. A brilliant and simple solution was proposed, motivated by Grand Unified Theories. The GUT scalar field fueled the superfast expansion that solved all the problems of the Big Bang model, a compelling notion linking the physics of the very small with the physics of the very large. If proven right, it would be a decisive step toward the Final Truth. However, it was soon shown that the original GUT-motivated approach to inflation didn't work: such a universe would never transition back into the usual Big Bang.[15] The whole GUT enterprise remains speculative to this day. As a way out, models stripped inflation to its bare minimum: a hypothetical scalar field displaced from its equilibrium state, like a ball rolling downhill, and filling a large enough patch of the early cosmos to make it expand with superluminal speed. The elegance that remains is the simplicity of the inflationary idea. The scalar field that fueled inflation remains elusive, a theoretical construction that may or not exist. Still, the idea solves so many cosmological problems that something like it must have happened. It just need not have anything to do with unification.

What we have learned this far is that we can solve most of the issues with the Big Bang model with hypothetical scalar fields whose origin remain a total mystery. Unifiers will quickly state that these fields are surely a consequence of a yet unknown true unified field theory. After all, superstring models abound with scalar fields of all kinds. Or maybe GUT theories do work, and we just don't know how yet. Maybe so, but *maybe* is not a very scientific word. We have no observational evidence for GUT unification, even though its first version was proposed more than thirty-five years ago. The same is true for superstring models. In fact it's worse there, as we don't even know how to look for evidence specific to superstring theories, apart from some very particular cases. The small patch of space that inflated to become our Universe may have done so fueled from a source that may or may not have been a scalar field. Unless we can detect fossils left from this era, and there are a few possibilities, we may never know. And even if we do, the data may not give us enough details to differentiate between models. We should, of course, keep trying, but we should also keep in mind that we may be permanently blind to the details of the cosmic birth. Perhaps the Universe is trying to tell us something about our dreams of a final theory. Perhaps we live in a cosmos much more mundane and imperfect than our expectations of sublime explanations and high symmetry make us believe. Perhaps all that is needed is a small patch of weirdness.

24 DARKNESS FALLS

Looking back, we can see how much the Universe, or more exactly, our view of it, has changed over the past few hundred years. For Columbus, Earth was fixed in the center of everything and the cosmos was bound by the (crystal) sphere of the stars. God reigned supreme over all, a constant presence in people's lives. For Benjamin Franklin, the Sun was the center, and the solar system ended in Uranus, discovered in 1781, nine years before his death. Neptune was not yet known. God was the creator of the Universe and its laws but abstained from interfering in world affairs. Einstein, in 1917, made use of his brand-new general theory of relativity to propose the first modern cosmological model. In the absence of data, he used Ockham's razor and assumed that the Universe was static and spherical. He reluctantly changed his mind in 1931, after learning of Hubble's discovery of receding galaxies, conceding that the cosmos need not be static. God was an abstraction, a metaphor for Nature's mathematical order, accessible to the human mind. The first Moon landing happened in an expanding cosmos, with an age estimated between a few billion and over 20 billion years. The matter composition was believed to be the ninety-four naturally occurring elements of the periodic table. There were also neutrinos and photons filling up the cosmos, both believed to be massless particles. In the four decades since Neil Armstrong's epic moonwalk, our view of the cosmos has changed dramatically.

Already in the early 1930s there were intimations that some of the matter in outer space is invisible. Its presence is inferred by its gravitational tug on stars and galaxies, that is, on the familiar kind of matter that bundles up into shiny objects we can see. The same way that the Sun's gravitational pull causes the planets to orbit around it, galaxies appeared to move about each other in ways that could only be explained if some kind of invisible matter was pulling them around.*

In the early 1970s, astronomers Vera Rubin and W. K. Ford demonstrated that even within an individual galaxy, stars move in ways that could only be explained if there were large amounts of invisible matter. This sounds more mysterious than it is. We, for one, are invisible. Even though some of us may "glow" or be "brilliant," we don't produce electromagnetic radiation within the visible range, limiting our emissions to invisible frequencies, such as infrared. A person would be hard to spot in a dark field. The same with planets and moons. They don't produce their own light, only reflect that of their main star.[16] Accordingly, the most natural candidates for tugging galaxies around were invisible objects made of ordinary matter, such as low-mass stars that can't shine or large cold clouds of hydrogen sprinkled with heavier chemicals. Presumably, neutrinos could also contribute, as they are supposed to swarm around galaxies. To the surprise of many and the delight of many others, things turned out not to be so simple.

Invisible matter is known in astronomy as *dark matter*. The surprise was not so much that it exists, but that it cannot be made of ordinary matter. Quite simply, there isn't enough normal stuff around to explain either how galaxies spin about themselves or around each other in clusters. But if dark matter is not made of ordinary protons and electrons, what *is* it made of? Faced with this dilemma, some physicists attempted to explain away the existence of dark matter by attributing the strange behavior of galaxies to limitations in our understanding of gravity. Maybe, they conjectured, Einstein's general theory of relativity acts differently at large distances. Ordinary matter is all there is and we simply need a new theory of gravity. Many ideas have been put forward, although they remain mostly unconvincing. A new theory of gravity needs a solid conceptual foundation, and these models offer none. Furthermore, there is one observational effect of dark matter that cannot be properly explained by any modification of gravity proposed to date.

*Note that the planets also pull on the Sun: the gravitational force goes both ways, as described in Newton's third law of motion, the one that states that for every action there is an equal and opposite reaction. However, due to the Sun's much larger mass, it barely responds to the planets' combined tugging.

According to Einstein's theory, matter bends space. Therefore, if light from a faraway source passes through a region of bent space it will curve and distort in ways similar to what happens when it passes through a glass lens. This phenomenon, known as gravitational lensing, is a beautiful consequence of Einstein's theory. Its importance for dark matter is that the net light bending is only sensitive to the total amount of matter and to how it is distributed in space, not to whether it shines or not. So, looking at images of nearby galaxy clusters illuminated from distant sources, astronomers can see the bending and distortions of light as it passes by and, from them, compute the amount and distribution of *all* matter in the cluster, visible and invisible. The observations, which are quite beautiful to see, leave no doubt that the amount of matter in the clusters and in individual galaxies is far larger than that in shining matter.[17] The best current estimates place the ratio of normal to dark matter as a little under 1:6. That is, there is roughly six times more dark matter than normal matter in the Universe. The problem, of course, is that no one knows what it is made of.

Many dark-matter candidates have been proposed over the past three decades. Small black holes, stars made of exotic states of matter, or, as some colleagues and I have proposed, stars made of scalar fields, known as boson stars, are examples of dark exotic astronomical objects. So far, none of these has been validated. The more popular candidates, however, are not large, starlike objects, but subatomic particles that do not interact via electromagnetism (so, no electric charge) or via the strong nuclear force. That is, they are not electrons, protons, or neutrons. Such particles are predicted to exist in extensions of the Standard Model, although none has been observed. If dark matter is indeed made of subatomic particles, they should be passing through us in large numbers and could, with some effort, be detected. At the time of writing, no detection has been confirmed, although the search goes on. Meanwhile, the possible existence of dark matter has inspired some wonderful works of fiction, such as Philip Pullman's *His Dark Materials* trilogy.

The compelling observational evidence for some kind of exotic dark matter shows that our Universe is mostly made of matter different from ours. There is great hope that we will find the composition of dark matter within the next decade or two. I hope we do.

Whatever it is, we owe dark matter our existence. Without it, the first stars and galaxies would not have formed during the first billion or so years of cosmic history. Everything starts with inflation. As with any field, the one responsible for inflation also fluctuated due to quantum uncertainties. The quantum jitters inherent in all of matter would have caused the scalar field to jiggle a bit like jelly: its values in space would fluctuate about a mean, like choppy waves on a lake. These initially tiny fluctuations in the values of the field were dramatically amplified during inflation, growing to astronomical sizes in different regions of the Universe: choppy waves became tidal waves. Since the field carries energy, these regions of space would have a lot of energy in them. Hundreds of thousands of years later, these big lumps of scalar field gravitationally attracted dark-matter particles, so that they swarmed around them in massive clouds. In turn, the gravitational pull from these large dark-matter clouds bent space around them. As water collects in puddles during a rainstorm, these space puddles eventually collected the protons and electrons that condensed into the first stars and galaxies. After burning furiously for a short time, these first stars departed with violent explosions, which triggered the birth of new stars. A few billion years later, this dance of creation and destruction led a particular gas nebula to collapse, forming our Sun and planetary system. We may be made of ordinary protons and electrons, but our origins are ultimately linked to dark matter and to the quantum fluctuations during inflation.

If I were writing this book fifteen years ago, I would end this portion on cosmology here. However, as if to prove once again how fluid our views of the cosmos are, everything changed in the late 1990s. The big surprise? Dark matter is not even the most important kind of darkness cloaking the cosmos.

25 DARKNESS RULES

In 1998, the physics and astronomy communities were taken by storm. Two groups of astronomers, one led by Saul Perlmutter, of the Lawrence Berkeley National Laboratory in California, the other by Adam Riess, then of the Mount Stromlo Observatory in Australia, observed supernovae located in galaxies billions of light-years away to study how fast they were receding. Their results were completely unexpected: the Universe is not only expanding, its expansion rate is *accelerating*. Again. Just as in inflationary cosmology, some kind of negative pressure is stretching the fabric of space faster than the speed of light, carrying the galaxies along. Even more remarkably, the acceleration seems to have started at a certain point in the past, roughly 5 billion years ago. The unusual observations were showered with intense skepticism and scrutiny. I must confess to having been intensely suspicious of the whole thing. There have been many false alarms in the history of science, amazing discoveries that were proven wrong, like cold fusion. An accelerating cosmos is too bizarre, too arbitrary to be true. Surely Nature couldn't be this capricious. But a decade has passed and the acceleration has survived, confirmed by other means, including its effects on the cosmic microwave background. The culprit has been called *dark energy*. The word *energy* differentiates it from dark matter, which is composed of localized material objects, be they subatomic or astrophysical. Dark energy is widespread, formless, and homogeneous, very much like an aether. We go through it, the solar system goes through it, the galaxy goes through it as if it were a ghost. Only at the largest distances, comparable to the size of the cosmos itself, can its presence be felt.

Myriad questions emerge: What is making the cosmos accelerate? What sets its measured value? Why did it ignite at a specific time? We don't have any answers. Inspired by inflation, the other period of acceleration in cosmic history, some say that dark energy is an unusual kind of scalar field, called, in a bow to Aristotle, "quintessence," the fifth essence.* Others say that the aetherlike substance is the net result of the intrinsic quantum jitters of empty space, the bubbling up and crushing down of countless particles from the vacuum, predicted from the uncertainties of quantum mechanics. If this explanation is correct, and it quite possibly is, phenomena at the smallest of scales determine the fate of the whole cosmos, a beautiful wedding of the physics of the very small and that of the very large. Of course, just as with dark matter, it is always possible that our theory of gravity does fail at cosmic distances and the acceleration is explained with a new theory. At this point nothing is certain, although observations are steadily closing the window on models of quintessence. What we do know is that dark energy makes up about 73 percent of the stuff in the cosmos. The rest is made of dark matter (23 percent) and our humble protons and electrons (4 percent). The shocking revelation of modern cosmology is that the Universe is 96 percent unknown. The more we know the more there is to know.

I am confident that the present dark ages are a prelude to a new age of discovery. In time, we should be able to solve the mysteries of dark matter and of dark energy. It may be quite a while, however, especially for dark energy. Physicist Leonard Susskind, in a somewhat disconsolate tone that summarizes the feelings of many, mused, "We could be wrong about cosmology for the next thousand years. Deeply wrong."[18] I would counter that we have *always* been "wrong" about cosmology, and will always be. There is no final "right" to be arrived at, only a sequence of improved descriptions of the cosmos. Each era, each generation even, will describe the Universe in ways that may be radically different from the preceding one. Future cosmologists no doubt will see our present description of a cosmos engulfed in dark materials with the same amusement that we see the Earth-centered cosmos of Columbus. Hopefully, they

* I imagine that the other four essences are the fundamental interactions: gravity, electromagnetism, and the strong and weak nuclear forces. Bad naming, as the hypothetical scalar field is not a fundamental force, and forces are not essences. The other four essences could be the other kinds of stuff filling up the cosmos: dark matter, photons, protons and neutrons (aka baryons, see Part III), and neutrinos and electrons (aka leptons).

won't feel superior to us, just aware of being better equipped with tools and theory. If they are wise, as we hope they will be, they will also be grateful for our youthful excursions into Nature's workings.

The past few decades have confirmed yet again that our view of the cosmos, as is the cosmos itself, is in constant flux. By now we should have learned that it is futile to hold on too strongly to our current description of the Universe. For there is no question that it will change. New technologies will force it to. What we make of the Universe is a reflection of who we are. And who we are, the way we see the world around us and ourselves, changes as our tools of exploration change. Even as we succeed in creating an impartial science, free of subjective judgment and sharing a language accessible to all, the Universe will always be a human construction. The Universe is what we measure.

We now live in an accelerating Universe, with its three main material components balanced quite finely to produce a flat overall geometry. This equilibrium is as fragile as a pin standing up. A previous era of accelerated expansion, inflation, has been proposed to address some of the limitations of the Big Bang model, including why the observed Universe is so flat. Stripped to its bare minimum, inflation seems to require only a small patch of weirdness, a region of the young cosmos with enough energy and (negative) pressure to produce an accelerated expansion. The original motivation for the theory of inflation—that the scalar field responsible for grand unification drove the faster-than-light stretching of space—failed. There is absolutely no direct correlation between inflation and unification. At this point, it is just a phenomenological description based on the possible existence of a scalar field (or fields) capable of feeding an accelerated cosmic expansion. Countless models were proposed in its place. Even superstring theories, which were proposed as unifying schemes, lose their appeal in the vastness of the "superstring landscape"; in numbers absurdly large anything is possible, including our improbable Universe, this rare bubble that emerged from the quantum multiverse with the right amount of inflation to be flat, with the right amount of dark matter to trigger galaxy formation and the birth of the stars, and with the right amount of dark energy to drive the current expansion.

Some of my colleagues vehemently argue against such probabilistic explanations. Echoing Einstein, they believe in an underlying order, a deterministic explanation for all there is. Having started as a Unifier, I too struggled with feelings of powerlessness, at the possibility that not all is explainable. However, as we have done with quantum mechanics, it's time to let go. In the same way that quantum indeterminacy will always limit our understanding of the world of atoms and particles, our limited knowledge of the physical world forbids the construction of a Theory of Everything. The very notion of such a theory is nonsense, as it presupposes that we know all there is to know and thus to unify. But unify what? If we can never know all there is to know, we will always have an element of uncertainty about the natural world. There is no final unification to be attained, only better models to describe the physical reality we can measure. Even as we improve our tools and increase our knowledge, we also expand the base of our ignorance: the farther we can see the more there is to see. As a consequence, it is impossible to contemplate a point in history when we will know all there is to know. The uncertainty of knowledge is as permanent as quantum uncertainty. Hard as this may be to accept, it is a fundamental limitation of human understanding. Only our intellectual vanity precludes us from seeing this clearly and moving on. Science will not be diminished in its grandiose task of explaining Nature if it doesn't have a unified dream to pursue.

Even if superstring theories one day prove relevant to explain our Universe and do what they are set out to do, they will not be the final word. Our explanations of Nature are never final—they only grow more effective at explaining increasingly accurate data. They are narratives we construct with our remarkable ability both to make tools and to make sense of what they measure. Given the discoveries of the past few decades, we need only open our eyes to see that our narrative is headed in a novel direction: it is not that a special Universe produced special beings; it is that a *mundane* Universe produced special beings.

What we are learning at an ever-increasing rate is that there is no grand plan, no cosmic blueprint, no overarching explanation for our Universe. It is here as it could not be, a cosmos that emerged from a bubble that accidentally had the ingredients to survive its own collapse and to promote, through the interactions among its material components, the increasing complexification of form that eventually led to living beings. The order that we assign to Nature is the order we seek in ourselves. The world is only beautiful because we think about it.

PART III

The Asymmetry of Matter

26 SYMMETRY AND BEAUTY

Ask any physicist about symmetry and the answer will probably be a variation of the following: "Nature is symmetric. Through mathematics, we uncover these symmetries. Our equations and theories are the embodiment of this inherent order of all things. Symmetry is beautiful and beauty is truth." A Unifier would add that "there is an overarching symmetry that encapsulates natural phenomena at the most fundamental level: the elementary particles of matter and their interactions are manifestations of this one symmetry, expressed in a unified field theory. Find it and we will understand Nature at its deepest level. The unified theory is the ultimate truth."

There is no question that we crave symmetry. We surround ourselves with objects of symmetric design: computers, dinner plates, cars, chairs. We look at each other and see it: a person with the left eye half an inch below the right is grotesque and unattractive. Symmetry and order have been reflected in our depictions of the divine even before the advent of monotheistic religions. In religious iconography, gods and angels are portrayed as beautiful; their faces are perfectly symmetric. Demons are ugly; their faces are distorted. At a more mundane level, something as mild and harmless as a large mole spoils the overall balance of a face. To teenagers, a pimple may incite despair.

Left-handed people, a definite minority at about 10 percent or so of the population, are traditionally considered more prone to lunatic behavior. In Italian, the left is *la sinistra,* which is related to *sinister.* Contrast with its opposite, *la destra,* or English *dexter.* Think dexterity. In grade school, my teacher, who looked terrifyingly like the Wicked Witch of the West, forced me to write with my right hand, making me feel like a freak. "Your horrible handwriting, that's *not normal,*" she would yell on and on. In spite of her bullying, I'm proud to say that I remained left-handed.

Throughout the ages, misfits, monsters, and any creatures with physical deformities have been shunned, and exhibited as curiosities at princely courts, popular fairs, and the circus. Believed to be bad omens, they incited fear and violence. In the Middle Ages, the birth of a two-headed calf or of conjoined twins, unusual and grotesque, was seen as a sign of difficult times ahead. If it happened near the end of a century, it heralded nothing less than the upcoming apocalypse.

We have been evolutionarily programmed to seek the beautiful, the harmonious, the well proportioned. Females with a waist-to-hip ratio of about 70 percent are the most fertile due to an ideal balance of progesterone. The Venus figurines of the Paleolithic—sculpted more than twenty-eight thousand years ago, Rubens's plumpy Renaissance models, and even 1960s skinny model Twiggy, all considered, in their own time, highly desirable, satisfy that same approximate ratio.

A hunter in a forest is acutely aware of the spatial patterns surrounding him. His survival depends on it. The presence of a predator or of an enemy warrior, both disruptions of the usual scenery, were reason for alarm. The periodic motions of celestial objects, such as days following nights or the return of the seasons, are regular patterns in time. Comets, eclipses, meteorites, and other disruptions of periodicity used to be feared, interpreted as messages from unhappy gods ready to exact punishment on the helpless humans below. From very early on, order was equated with safety, and symmetry with the predictable.

In art and in music, what is considered modern is often asymmetric. During the early twentieth century, the break with classic art was, to a large extent, expressed as the breaking of symmetric patterns in Nature and the human figure. Picasso's women are still considered "ugly" at first viewing. In music, atonality marked a break with the old aesthetic that embraced harmony at its core. Arnold Schoenberg's and Alban Berg's atonal compositions sound bizarre and, to many, unbearable. The 1913 performance of Berg's *Five Songs on Picture Postcard Texts by Peter Altenberg*, conducted by

Schoenberg in Vienna, ended in a riot. Igor Stravinsky got his share of booing for lesser sins.

Asymmetry is uncomfortable. It exposes deep fears, some of them long forgotten, their roots reaching back to the monotheistic religious tradition. An asymmetric world cannot possibly be the work of a perfect God; if the world is asymmetric it must be godless. And if the world is godless, we are alone, left to handle our own problems, left to deal with our predators and enemies, with our bad choices and losses. An asymmetric world is scary. Humans yearn for the protective embrace of the symmetric, the ordered.

Around 580 B.C.E., the Greek pre-Socratic philosopher Anaximander based the first-ever mechanical model of the cosmos on the most symmetric of all shapes, the circle. As we have seen, circles dominated astronomy until 1609, when Kepler showed that Mars's orbit was elliptical. The shift, even for Kepler, was not easy. He struggled with self-doubt for years before accepting his groundbreaking conclusion. It would not have been his preference, but he knew Tycho's precise data couldn't be ignored. Undaunted, to the day of his death Kepler remained certain that the perfection of the cosmos, reflecting that of God, was hidden at a deeper layer of reality. He just had to dig deeper.

Scientists only abandon symmetry with great reluctance. There are at least two very good reasons for this. The first is that symmetry has proven to be a phenomenally efficient tool to describe Nature. From cosmology to particle physics, much of what we have learned depends on different kinds of symmetries. The crystalline structure of many solids, from kitchen salt to diamonds, is absolutely critical to understanding their properties. Symmetric systems are easier to describe mathematically; symmetric equations are easier to solve. Sometimes, imposing a specific type of symmetry generates predictions that are spectacularly confirmed by experiments. It is as if thought preceded Nature, as if aesthetics was deeply built into the natural order. This is particularly true in particle physics: by imposing mathematical symmetries on the theories that describe how particles of matter interact with each other, physicists were able to predict the existence of *new* particles, previously unknown to exist. We will soon see some examples of this.

The other reason for our long love affair with symmetry is that it's beautiful. Now, beauty is not easy to define, even within a more precise scientific context. A beautiful theory is both simple and powerful: relying on a minimal number of assumptions it explains a wide range of phenomena. A beautiful theory is also the nearest to perfection: remove a small piece and the whole edifice crumbles. As Weinberg wrote in his *Dreams of a Final Theory:* "The beauty that we find in physical theories like general relativity or the standard model [of particle physics] is very [much] like the beauty conferred on some works of art by the sense of inevitability that they give us—the sense that one would not want to change a note or a brush stroke or a line."[1]

It is very important for the reader to understand that this book is *not* a manifesto against symmetry. That would be both foolish and incorrect. Symmetry is, and will remain, a key ingredient of our theories. Symmetry is beautiful, and, in the sense that I defined above, our physical theories are beautiful. Many structures in Nature *are* indeed symmetric. One has to simply look at a sunflower, an ant, or a quartz crystal to appreciate this. The problem starts when symmetry is taken too far and is enthroned as dogma. Symmetry is beautiful, but, apologies to John Keats, beauty is not necessarily truth. Or, for that matter, truth beauty.

We have seen how our Universe seems to be the result of a random quantum fluctuation that burst out of the vacuum some 14 billion years ago. We have also seen how dark materials of an unknown nature surround us, and how these materials are linked to both the birth of galaxies and the ultimate fate of the cosmos. We do not yet know what they are or why they appear in the quantities we measure. In fact, dark energy, the diffuse medium that permeates the cosmos, is downright mysterious. Whether we want it or not, this is our Universe. Different from what it was fifty years ago and different from what it will be fifty years from now. To be sure, there is plenty of beauty in it. But it's another kind of beauty, a beauty of Becoming, not of Being; of change and transformation, not of equilibrium and stasis; of imperfection, not perfection. There is a need for a new kind of aesthetic in science, a new kind of beauty that leaves behind expectations of order and symmetry inspired by age-old monotheistic beliefs. This new aesthetic is founded on a single principle: the realization that Nature creates from imbalance.

27 A MORE INTIMATE LOOK AT SYMMETRY

Human faces, cars, wineglasses, balls, CDs, all of these are symmetric. The reader knows this instinctively. Each has one or more symmetries that can be expressed mathematically. They are examples of *spatial symmetries*, in that they are related to spatial perception, to distance measurements and proportions. To properly understand symmetry we need two ingredients: the symmetry operation and the object operated upon. A ball, if perfectly clear of any blemishes, is highly symmetric: turn it around and it looks exactly the same. In this example, the ball is the object and the symmetry operation is a rotation of the ball around its center. A human face is what we call left-right or reflection-symmetric: one side is identical to the other. Well, almost. Human faces are only approximately symmetric. We all know that the two sides of our faces are not exactly alike. Something is always there to break the perfect symmetry: a scar, a mole, a wrinkle, the parting of the hair. A better example of a reflection-symmetric object is a bicycle, if viewed from the front or back. The symmetry operation takes every point on one side to another on the opposite side at exactly the same distance from the center. If the object is reflection-symmetric, the two sides will look exactly the same.

Plato was highly suspicious of perfect symmetries in the real world. He thought them impossible. Only in the world of the mind could perfection exist. A circle is only perfect as an idea. Any concrete representation of a circle will never be as perfect as the idea of it. When combined with the Pythagorean number mysticism, this Platonic idealism reaches an explosive conclusion: mathematical relations between forms and numbers hide the secret to the ultimate nature of reality. The hidden code of Nature is written in the language of perfect forms. The entry password into this realm is *symmetry*. This idea, or better, ideal, percolates through the whole of the physical sciences. Every student of physics and chemistry starts by studying symmetric systems: cylinders, circles, spheres. When describing an object or a physical system (a molecule made of two or more different atoms, stars revolving around each other) that is only approximately symmetric we treat it as one with small distortions (or perturbations) about perfect symmetry, hoping that the perturbations remain small as time goes by. If they do, all is well. If they don't, things get pretty complicated.

Physics is the art of approximations. Only a very small number of physical problems have easy solutions due to their high level of symmetry. We are good at solving oscillatory motion (so long as the oscillations are small) and we know how to use quantum mechanics to solve for the energy states of hydrogen, the simplest of all atoms.[2] Complex oscillations and multi-electron atoms are solved (when possible) via sophisticated approximation methods that usually rely on small perturbations about the symmetric solution or state(s). Of course, computers have changed all this dramatically. Problems that couldn't be solved a few years ago by the best mathematicians are now solvable on a laptop by first-year students. To a large extent, analytical prowess has given way to programming prowess. Laplace's supermind is still not at hand, but our silicon cousins have expanded our minds considerably. In principle, computers can handle arbitrarily large perturbations and highly asymmetric shapes and systems. In practice, machines and their programmers have their limitations. Even so, thanks to computers, spatially asymmetric systems no longer represent an insurmountable barrier. Asymmetry has entered the mainstream.

In what follows, I will often refer to "external" and "internal" symmetries. It's now time to clarify the distinction. External symmetries manifest themselves in space and time. I mentioned several spatially symmetric shapes, such as balls, cars, and CDs. A square is also highly symmetric. Turn it by 90 degrees around its center and it looks the same. That's a *rotational* symmetry. A square is also reflection-symmetric: one half is the mirror image of the other. A good way to think of symmetry is as change without change: after the symmetry operation is performed, the object remains the same, as if nothing had happened.

In particle physics, symmetries are subtler. There are, of course, external space and time symmetries. As long as the same conditions are maintained (altitude, temperature, etc.) an experiment will give the same results if performed in the United States or in Brazil, on a Monday or on a Saturday. Also, it doesn't matter if your laboratory is oriented north or west (unless Earth's magnetism can affect the results). One of the deepest consequences of symmetries of any kind is their relationship with conservation laws. Every symmetry in a physical system, be it balls rolling down planes, cars moving on roads, planets orbiting the Sun, a photon hitting an electron, or the expanding Universe, is related to a conserved quantity, a quantity that remains unchanged in the course of time. In particular, external (spatial and temporal) symmetries are related to the conservation of momentum and energy, respectively: the total energy and momentum of a system that is temporally and spatially symmetric remains unchanged.*

The elementary particles of matter live in a reality very different from ours. The signature property of their world is *change:* particles can morph into one another, changing their identities. These changes can happen spontaneously, as when an isolated neutron decays into a proton, an electron, and an antineutrino (more on this "anti" soon). Particles can also change identities when they collide with one another, as when a proton traveling from the Sun hits the air molecules on Earth's upper atmosphere, or when experiments in particle accelerators force protons to hit other protons. The goal of particle physics is to uncover the rules controlling these changes. Said another way, one hundred years of research has established that particles interact and morph into one another according to a set of strict conservation principles, each one of them related to a symmetry of the external (related to space and/or time) or the "internal"

*The momentum of an object, for those who haven't looked at these things in a while, is the product of its mass with its velocity. Unlike energy, a scalar quantity, the momentum is a vector, having a direction. Energy may change nature (say, from chemical to electric) while momentum may be exchanged (say, via collisions between a system's parts); but in a conservative system, their total amounts remain the same.

kind. While external symmetries dictate how energy and momentum flow during particle interactions, internal symmetries dictate how particles can change identity. A particle has the potential to become many others, as if it suffered from multiple personality disorder. Which specific identity will take hold depends on the particular circumstances of its interactions. For example, a lonely neutron can morph into a proton, an electron, and an antineutrino. But when the neutron is part of a stable atomic nucleus, that is, when it interacts with other protons and neutrons in a stable environment, it retains its neutron identity. One of the greatest triumphs of twentieth-century particle physics was the discovery of the rules dictating the many metamorphoses of matter particles and the symmetry principles behind them. One of its greatest surprises was the realization that some of the symmetries are violated and that these violations have very deep consequences.

28 ENERGY FLOWS, MATTER DANCES

The history of particle physics is one of ongoing revelations. The first was that atoms are actually not indivisible, which Leucippus and Democritus had supposed in pre-Socratic times, but rather that they are made of smaller constituents. In 1897, J. J. Thomson announced the discovery of the electron, the first elementary particle of matter to be identified. Thomson showed that different chemical elements have electrons, that all of these electrons have the same mass and negative electric charge, and that they are much lighter than the lightest of all chemical elements, hydrogen. He conjectured that electrons were constituents of all atoms. That being the case, how could chemical elements—being made of the same ingredient—be so different from one another?

In Thomson's time scientists knew that each chemical element had a different weight. Many years earlier, in 1869, Dmitry Mendeleyev proposed his first version of the periodic table, arranging the known chemical elements according to their increasing atomic weight. He found that elements with similar chemical properties (such as metals or the alkaline earths) fell naturally into periodic patterns, called families. Mendeleyev's faith in the regularity of Nature brought him fame. Noticing that there were holes in the table where missing elements with well-determined properties should be—such as atomic weight and chemical affinity—he predicted that they *had* to exist. Within a very short time, germanium, gallium, and scandium were discovered with precisely those properties Mendeleyev had predicted. This sort of reasoning by regularity or periodicity plays a major role in our understanding of matter.*

Thomson's discovery of the electron started a revolution in our understanding of the atom. In 1918, Ernest Rutherford conjectured that the hydrogen nucleus was made of a single particle, the proton, with positive electric charge exactly opposite that of the electron and with a mass about two thousand times larger. According to this (correct) model, atoms are electrically neutral with mass mostly located in the nucleus. In 1932, James Chadwick discovered the neutron, the last big player in atomic models. These results led to a startling simplification: the more than one hundred naturally and artificially occurring known elements collected in the periodic table are combinations of just three matter particles: protons, neutrons, and electrons.

The number of protons in the nucleus sets the identity of a chemical element. Variations called *isotopes* may occur due to the number of neutrons. For example, hydrogen, having one proton, exists in three forms: normal hydrogen, deuterium, and tritium. Each has a single proton in the nucleus. But deuterium and tritium, which occur in much lower abundances, have one and two neutrons in their nuclei, respectively. They are both isotopes of hydrogen.

During the 1920s and '30s, experiments further revealed the alchemical fluidity of nuclear matter: elements transform into one another either after being bombarded with particles or spontaneously, through radioactive decay. Radioactivity is nothing more than the emission of particles from an atomic nucleus. If the uranium isotope U-238 (92 protons and 146 neutrons) emits an alpha particle (a nucleus of helium, with 2 protons and 2 neutrons) it becomes an isotope of the element thorium, Th-234 (90 protons and 144 neutrons). In symbols, U-238 $\rightarrow$ Th-234 + alpha particle. Such natural alchemical transformations conserve energy and momentum. And they also conserve electric charge: the total electric charge before and after is always the same. (Just compare the total number of protons on

* Strictly speaking, Mendeleyev's ordering is incorrect. The elements should be grouped by their *atomic number*—the number of protons in their nuclei—and not by their *atomic mass*—the sum of protons and neutrons in their nuclei. Neutrons are slightly heavier than protons, and this causes issues with the ordering of some elements.

the left and on the right of the reaction.) Charge conservation is an example of internal symmetry. Any transformation violating such laws is "forbidden," that is, not observed. With further experimentation and mounting evidence, a small number of conservation laws and related symmetry principles were seen to describe the transformations of matter and energy. Material objects became secondary to the rules determining their behavior.

Four years before the discovery of the neutron, Paul Dirac attempted to marry quantum mechanics with special relativity. If the electron moved about the nucleus, it would do so at near-relativistic speeds; there should be corrections, even if small, to the predictions of quantum mechanics. Furthermore, Schrödinger's quantum mechanics didn't include a key property of the electron, its intrinsic rotation, or spin. It had to be added by hand. To his surprise, Dirac found two solutions to his relativistic equations, with opposite electric charges. One described the negatively charged spinning electron, to be sure. But the other . . . his first guess was that it described the proton. J. Robert Oppenheimer, who later led the Manhattan Project during World War II and is one of the most tragic figures in modern history, quickly realized that the other spinning solution described a "positive electron." The predicted new particle was called *positron,* that is, an electron with a positive electric charge. In 1932, the same year Chadwick discovered the neutron, American physicist Carl Anderson, apparently unaware of Dirac's and Oppenheimer's predictions, also discovered the positively charged electron.

Detailed observations determined that positrons don't appear naturally on Earth: they result from decays and collisions of more ordinary particles. Their source is the Sun, protons routinely spewed out as part of its heat-churning processes. After traveling 150 million kilometers, such *cosmic ray* protons arrive at Earth's upper atmosphere. They hit a nucleus, say of nitrogen, and produce many particles. Those hit more nuclei and make more particles. Countless collisions create all sorts of secondary products, often called showers. Among them is the positron Anderson detected. These chain-like processes are the embodiment of Einstein's $E=mc^2$ relation: the energy of motion of the incoming protons (their kinetic energy) is literally transformed into new particles of matter and into photons. Energy flows and matter dances.

One of the ways to create a positron, in particular, starts with a high-energy gamma-ray photon hitting a proton; the photon's energy is enough to give the proton a major kick and to create, here it comes, an *electron-positron pair*. In symbols, we write the reaction as follows:

$$\text{photon} + \text{proton} \leftrightarrows \text{electron} + \text{positron} + \text{proton}$$

One sees the fluidity of matter in all its beauty: massless radiation (the photon) goes into massive particles. The two arrows indicate that the reverse is also possible: electrons and positrons annihilate each other into a puff of gamma-ray radiation (highly energetic photons). One also sees why, when we try to understand what goes on in the world of subatomic particles, conservation laws are of the essence. In the reaction above, energy must be conserved. If the photon is not energetic enough, the proton gets a kick but no electron-positron pair is created.* Momentum conservation determines the directions in which the particles fly out. And charge conservation requires that the two sides of the reaction have the same total charge. If the left-hand side has one unit of positive charge (the proton), the right-hand side must as well. Hence if a proton appears on the right (as it does here), the extra products must add up to zero charge. This is precisely what happens, as electrons and positrons have equal and opposite electric charge. There has *never* been a single instance of a particle reaction that violated either energy or electric charge conservation. Within the accuracy of our measurements, we can state that these are laws of Nature. There is choreography behind the apparently random dance of matter creation and destruction.

*Since electrons and positrons have the same mass, the gamma-ray photon must have a minimal energy equivalent to twice the mass of the electron (times the square of the speed of light).

29

VIOLATION OF A BEAUTIFUL SYMMETRY

The positron is called the *antiparticle* of the electron, the first discovered example of *antimatter*. Dirac himself soon realized that his theory of relativistic quantum mechanics, that is, the theory combining quantum mechanics and special relativity, resulted in equations that could also describe protons and their antiparticle, the antiproton. The equations always have two solutions: one for particles of matter and another for particles of antimatter. As the positron has opposite electric charge to the electron, the antiproton has opposite electric charge to the proton. Dirac had shown that the existence of antimatter is an unavoidable consequence of bringing quantum mechanics and relativity together. Few examples of a mathematical structure capable of predicting objects that are subsequently found in Nature are so impressive. No wonder Dirac believed that only beautiful equations could be right. His equation for the relativistic electron surely is both. Its prediction is clear: every particle of matter has its antimatter companion. Most of their properties are the same, such as their mass and spin. If the particle is stable, its antiparticle is stable. This is the case with the electron and the positron. If it's unstable, as the neutron is, its antiparticle is unstable. Their lifetimes, the average time for an unstable particle to decay, are the same. We can see why Dirac tried so hard to establish the existence of magnetic monopoles, to bring electromagnetism to its full perfect symmetry. I imagine that a part of him believed that only then would Maxwell's theory be truly beautiful. However, in spite of his efforts, and those of many others, things didn't go according to plan. Nature, as usual, is more creative than that.

And it has a very clear preference: there is hardly any antimatter in the Universe. The antimatter world is a sort of duplicate of the matter world. But the copy is not perfect; the electric and magnetic properties of particles and of their antiparticles are reversed. Other properties, having to do with how particles and antiparticles interact at subnuclear distances, are reversed as well. These differences between the two are at the core of one of the greatest unsolved mysteries of modern particle physics, an all-important asymmetry of Nature. Even though matter and antimatter appear in equal footing on the equations describing relativistic particles, antimatter occurs only rarely. The antiparticles that we see are fabricated in collisions between matter particles, be they in cosmic rays or in particle accelerators. Some may be made in violent astrophysical events, for example when black holes swallow whole stars. Somehow, during its infancy, the cosmos selected matter over antimatter. *This imperfection is the single most important factor dictating our existence.* Had matter and antimatter coexisted in equal amounts during early cosmic history, they would have annihilated each other to such an extent that the Universe today would consist mostly of a bath of radiation. Life would not exist.

Before we move on to explore the roots of this fundamental asymmetry of Nature, we must establish that it is indeed prevalent throughout the cosmos and not just a local effect. Could other regions of the Universe be made of antimatter? Could there be antimatter galaxies, for example?

According to quantum mechanics, yes, it is perfectly possible for anti-atoms to exist and hence for galaxies to be entirely composed of antimatter. Antihydrogen has been synthesized in the laboratory.* Now, we know that the Moon is not made of antimatter; if it were, poor Neil Armstrong and the entire lunar landing module would have disintegrated at touchdown in a gigantic explosion. The same with most planets of the solar system and some of its moons; probes have been there and survived to send us information. We can infer that our whole galaxy is made of matter. Had there been much antimatter around, we would have detected the excess gamma rays

* An antihydrogen atom has a positron surrounding an antiproton nucleus. Antihelium would have two positrons surrounding two antiprotons and antineutrons in the nucleus. It has never been seen.

coming from collisions of stars and anti-stars or from antimatter interstellar dust clouds interacting with normal dust clouds.

Current gamma-ray detectors push the boundary of a matter-pure cosmos to about 65 million light-years from us, well beyond our local galactic neighborhood. How much farther can we go? I would say to the entire observable Universe. In the late 1980s, when I was a postdoctoral fellow at the Institute for Theoretical Physics at the University of California at Santa Barbara, I collaborated with David Cline from UCLA, Floyd Stecker from NASA Goddard Space Flight Center, and then–UCLA grad student Y. Gao on research aiming to answer this question. Could there be huge domains, or bubbles, containing only antimatter in the Universe? If there were, at the boundaries of these domains matter and antimatter would collide and annihilate, contributing substantially to the net extragalactic gamma-ray background. Using a theoretical model, we estimated the production of gamma rays at the annihilation boundaries, investigating how its intensity varied with the sizes and the thickness of the domains of matter and of antimatter. We then compared our results with known observations of the extragalactic gamma-ray background. Our conclusion was that no such domains could exist without having already been detected. We live in a cosmos of matter.

Big Bang cosmology further constrains the existence of antimatter. The reader may recall that one of the triumphs of the Big Bang model is the accurate prediction of the abundances of the lightest nuclei, from hydrogen to lithium-7 (three protons and four neutrons).* This remarkable result relies on a single parameter, the relative matter-antimatter asymmetry. Specifically, at the time of *primordial nucleosynthesis*, that is, at about a second after the Bang, to every billion-and-one particles of matter there should have been one billion particles of antimatter. It looks tinier than it is. Think that in about one gram of matter there are some trillion trillion atoms.† That's a thousand trillion more than a billion! So, the required amount of asymmetry says that in a sample of matter with a billion anti-atoms there would be a billion-and-one atoms; for a

* This prediction was discussed in Part II.

† More precisely, in 12 grams of carbon-12 atoms there are 6×10^{23} atoms, the famous Avogadro number.

sample with ten billion anti-atoms there would be a ten billion-and-ten atoms; for one with a hundred billion anti-atoms there would be a hundred billion-and-one-hundred atoms, and so forth. Following this logic, after the complete annihilation of one gram of anti-atoms (that is, about 10^{24} of them), about a thousand trillion atoms (10^{15}) would remain.

Back to the early cosmos: had there been an equal quantity of antimatter particles around, they would have annihilated the corresponding particles of matter and all that would be left would be lots of gamma-ray radiation and some leftover protons and antiprotons in equal amounts.[3] Definitely *not* our Universe. The tiny initial excess of matter particles is enough to explain the overwhelming excess of matter over antimatter in today's Universe. The existence of matter, the stuff we and everything else are made of, depends on a primordial imperfection, the matter-antimatter asymmetry.

Once the asymmetry is established, the obvious next step is to explain it. Why should there be one-billion-and-one particles of matter to one billion of antimatter? What processes in the early Universe could have caused such an imbalance? In order to examine possible answers to these questions, we must first explore in more detail the symmetries and asymmetries of particle physics, as they are at the core of our imperfect cosmos and are ultimately responsible for our existence. If the hurried reader knows about quarks, leptons, gluons, and the three weak gauge bosons, she or he may skip the next chapter. Otherwise stay with me and explore the amazing world of subatomic particles and their interactions.

30 THE MATERIAL WORLD

The decades following World War II saw a true revolution in the study of matter. Machines capable of smashing nuclei against nuclei, electrons against positrons, and protons against antiprotons at increasingly high energies revealed a rich and surprising structure, very different from what even the patriarchs of the quantum revolution could have envisaged. When energy flows, matter dances in amazing ways.

To start with, we now count four fundamental forces of Nature, the interactions between the particles of matter. At human scales, we are familiar with gravity and electromagnetism. The fact that we know about them has to do with their long range: they both fall as the inverse of the square of the distance. If we could double the distance between the Sun and the Earth, their mutual gravitational attraction would drop by a factor of four. The very important difference between gravity and electromagnetism, or actually, between gravity and everything else, is that it is the only interaction that is *always* attractive. A chunk of matter will always exert a gravitational attraction on other chunks. It can't be neutralized by countering positive "gravitational charges" with negative ones, as electricity can. This is why gravity is the only important interaction at the cosmic scale; while the others get neutralized or are negligibly weak, gravity adds up, atom by atom.

The other two forces act only at nuclear and subnuclear distances. The *strong force* keeps protons together in atomic nuclei despite their electric repulsion. As we have seen, it also glues the neutrons there and keeps them stable. During the 1950s, experiments produced enormous amounts of particles that interacted via the strong force, as protons and neutrons did. There were so many new particles that physicists started to despair: if elementary particles of matter kept being found, what was the point of calling them elementary? As this story has been told many times, I will be brief here.[4] To simplify things, physicists gave these new particles a common name: *hadron,* from the Greek for bulky. There were two types of hadrons: *baryons,* such as protons and neutrons, and *mesons,* such as pi-mesons (aka pions). In the 1930s, Japanese physicist Hideki Yukawa predicted that pions were responsible for the stability of the atomic nucleus: protons and neutrons interact by exchanging pions, somewhat like children throwing snowballs at each other. Yukawa conjectured that the mass of the pions determined the range of the intranuclear interaction: the heavier the particle transmitting the force, he reasoned, the more energy is needed to launch it and the shorter the range needed to be. As everyone knows, heavy snowballs are harder to throw far.

The proliferation of hadrons was worrisome. Was the whole enterprise of finding the elementary building blocks of matter a fanciful dream? I wonder how many particle physicists from the 1950s felt somewhat at a loss, blaming Thales and the rest of the pre-Socratic Unifiers for their predicament. (At least those who knew anything about the history of science.) Should they carry on under the spell of the Ionian Enchantment? Or had they been deeply misguided about the fundamental nature of the material world?

Enter Murray Gell-Mann and George Zweig. In 1964, they independently put forward a brilliant idea. What if, like atoms, composed of only three particles, hadrons are composed of only a few elementary constituents? Gell-Mann called his candidate particles *quarks,* while Zweig called his aces. Quark was the name that stuck. A few years earlier, Gell-Mann had realized that many of the new hadrons could be organized in groups of eight (octects) using their electric charge and a new property he called "strangeness." One can think of it as a different kind of charge that certain particles carry: just as we may carry many forms of identification—driver's license, Social Security card, passport—particles carry many IDs, called quantum numbers. The most familiar of these is the electric charge, associated with the electromagnetic interaction. Strangeness is just another kind of charge.

Gell-Mann's approach, called, with Buddhist inspiration, "the eightfold way," was tremendously successful. Echoing Mendeleyev's work on the periodic table, Gell-Mann realized that there was an empty spot in his octet classification. When the omega-minus particle was found in 1964 with the properties Gell-Mann's scheme predicted, strangeness was validated. Once again, Nature conceded to our request for symmetry.

Gell-Mann is a master pattern finder. He received the 1969 Nobel Prize "for his contributions and discoveries concerning the classification of elementary particles and their interactions." There was no direct mention of quarks. Even then, quarks were still not part of the game. For good reasons, the physics community wasn't entirely convinced they existed: as constituents of protons, they would necessarily have fractional electric charge, something a lot of people were reluctant to consider. Also, no one had ever seen a free quark. In his remarkable 1964 paper (only two pages long!) Gell-Mann wrote: "It is fun to speculate about the way quarks would behave if they were physical particles of finite mass."[5] This suggests that he considered quarks a real possibility, although some authors insist that he didn't, at least initially. The paper is unique in another way. Gell-Mann cited James Joyce's *Finnegan's Wake* as the reference for the peculiar word *quark*.

To make a long story short, quarks are now widely accepted as the constituents of hadrons. Baryons are made of three quarks and mesons of a quark and an antiquark, the antiparticles of quarks. There are six types (or *flavors*) of quarks: up, down, charm, strange, beauty, and top. For example, a proton is made of the triplet uud (*u* for up and *d* for down), while a neutron is made of the triplet udd. So, every atomic nucleus is made of up and down quarks. A neutral pion is made of an up and an anti-up quark, or a down and an anti-down quark. All mesons and baryons are unstable; that is, they decay spontaneously into other particles. The only exception is the proton. Unless, that is, grand unification is correct, something we will address soon.

At the level of quarks, the strong interaction is mediated by particles called *gluons*. They play a similar role to the one that photons do for electromagnetic interactions. However, gluons are sensitive to a different kind of charge that quarks alone carry, their *color*. The theory describing how quarks and gluons interact to make up the hadrons, a powerful example of symmetry in action, is called *quantum chromodynamics*, or QCD. There are three possible colors: red, green, and blue. Just as atoms must be electrically neutral, any hadron must be color-charge neutral: each of the three quarks making up baryons carries a different one of the three complementary colors (and so add up to "white"), while mesons have a color and an anticolor.

If quarks exist, are they seen as free particles? Here things get more interesting. Initially, Gell-Mann and others believed quarks could be detected, possibly in cosmic rays. No such luck. Quarks are imprisoned inside hadrons and cannot get out. This *confinement* (or color confinement) is one of the defining properties of quarks. You can't yank a quark out of a color-neutral meson or a baryon. If you try, you end up creating a quark-antiquark pair, that is, another meson. This is often compared to what happens when you break a magnet in two: you end up with two magnets, each with the usual two opposite poles. Loosely speaking, as you pull two quarks apart, the attraction between them grows, as if a kind of stretchy string connected them. The string is made out of gluons. If you keep applying more energy to force the quarks farther apart, the pair snaps into two pairs, just as a piece of string would snap into two pieces of string. The energy forcing the separation is converted into a new quark-antiquark pair, a meson.

At the other extreme, if you bring quarks closer together, they start to ignore each other and act as if they were free particles. Since probing short distances means using higher energies, this property of quarks only manifests itself in high-energy collisions. In 2004, David Gross, David Politzer, and Frank Wilczek received the Nobel Prize for developing the theory explaining this curious quark behavior, aptly called *asymptotic freedom*. This property is very important for understanding the young Universe. Recall that at early times matter gets squeezed into smaller and smaller volumes and the temperature increases. At about one-millionth of a second after the Bang, the temperature reaches values related to the masses of mesons and baryons: quarks and antiquarks break loose from their confinement and behave as free particles. During earlier times, the primordial soup contained quarks and gluons instead of hadrons.*

* Recall that, in cosmology, it's common practice to relate temperature to energy. The temperature of the universe is usually defined as that of the ubiquitous photons. High temperatures (early cosmological times) correspond to high energies and vice versa.

So much for the strong nuclear force. What about the weak force? Its claim to fame is that it is the force responsible for radioactivity. It actually can convert a down quark into an up quark and hence morph a neutron (udd) into a proton (uud). As with electromagnetism and the strong force, it acts via the exchange of its own three force carriers. With masses eighty to ninety times larger than the proton mass, these carriers are quite heavy. As a consequence, the weak interactions have very short range, limited to act within subnuclear distances. The three weak force carriers are called W^+, W^-, and Z^0, admittedly not too interesting. Sometimes they are referred to as the *weak gauge bosons*, even worse. Sheldon Glashow, Abdus Salam, and Steven Weinberg, whom we encountered earlier on, predicted their existence in the 1960s. Their discovery in the '80s was a great validation of our theoretical understanding of the fundamental forces. All three of them—electromagnetism and the strong and weak nuclear forces—are described in a very similar way, through the action of force carriers. The big difference between these—photons, gluons, and the weak gauge bosons—is that the photons do not interact with each other. This makes the strong and weak nuclear forces more complicated and, of course, richer.

Completing this crash course on the particles of matter and their interactions, we present the *lepton*, from the Greek for "light weight." Since leptons don't interact via the strong force, they are not part of atomic nuclei. There are six of them, the electron and the electron neutrino being the two most famous. The other leptons are the *muon* and its neutrino, and the *tau* and its neutrino. Note the pairing: each of the three negatively charged leptons has its (electrically neutral) sidekick neutrino. This means that when electrons interact via the weak force we expect to see electron neutrinos; when muons are involved, we should see muon neutrinos. In 1936, Carl Anderson, the same man who four years earlier had proved that antimatter existed, discovered the muon, a particle similar to electrons but about two hundred times heavier. Anderson found muons in cosmic rays, as by-products of upper-atmospheric collisions. The skies rain many invisible gifts upon us. Contrary to electrons, which are stable, muons decay in about a microsecond. A common decay path is muon → electron + electron-antineutrino + muon-neutrino. The tau lepton is even shorter lived, decaying after about a trillionth of a second. It's also quite massive, almost twice as much as a proton, making *lepton* a curious misnomer.

Putting all this information together, we learn that the particles making up ordinary atoms—protons, neutrons, and electrons—are the only stable ones. (Or at least extremely long-lived.) No wonder we don't commonly see things made of other hadrons and leptons. Their fleeting existence can only be captured with tools that greatly amplify our grasp of reality: powerful accelerators and their detectors.

Now that we have some familiarity with quarks, leptons, and their interactions we can use this knowledge to explore what really matters to us: the symmetries and asymmetries of particle physics.

31 SCIENCE OF THE GAPS

The collection of all the information on the elementary particles of matter and their interactions makes up the Standard Model of particle physics, mentioned before. In all, it is a spectacular achievement of the human intellect: not just in its devising of the theories that explain countless observations, but also in inventing the various technological tools that made the observations possible. To reduce the hundreds of particles of matter to only twelve (six quarks and six leptons) is an enormous simplification. No wonder Weinberg named one of the chapters of his *Dreams of a Final Theory* "Two Cheers for Reductionism." The quest to explain the material composition of the world started with the very first question asked in philosophy and is still very much with us today. Thales believed that all matter was reducible to a single constituent, water. The relevant point here is the nature of his answer and not the specifics. It betrays a deep belief in a unifying principle behind Nature's myriad materials. In other words, the first Ionian philosopher was a Unifier. And here we are, twenty-five centuries later, pursuing essentially the same goal. Should we keep searching for a unified description of matter? Is it indeed there, waiting to be discovered? Or is the seductive power of the Ionian Enchantment, bolstered by thousands of years of monotheistic culture, blinding us? Should we, perhaps, go beyond Isaiah Berlin's "Ionian Fallacy" and call it the "Ionian Delusion"?

An obvious answer is "Well, since we don't know, we must keep looking." No question about it. We must and should keep looking. Exploration and curiosity propel the advancement of knowledge. Many are plowing ahead, constructing ever more convoluted theories, trying to reach the presumptive final goal of reductionism, the unveiling of Nature's hidden code. We need brave explorers to set sail and risk everything for their quest. But when does the certainty of an existing endpoint to the search become a myth, an El Dorado? When does the drive for exploration become reckless obsession? Since the elusive final "truth" can always be pushed beyond the reach of experiments, the search may never end. The unification obsession may turn tragic with the refusal to accept that—even within the strict reductionist project of particle physics—a final theory is an impossibility since it presumes having complete knowledge of the natural world up to the smallest distance scales. We only know what we measure, and we can't measure it all. Paraphrasing Einstein (who was paraphrasing Kant), theory without experiment is blind, and experiment without theory is lame.

We have reached a peculiar situation in physics whereby experiments may confirm a theory but never refute it. For example, imagine that a given theory predicts the existence of a new particle. The details of the prediction depend on a parameter that can be adjusted. Reasonable arguments suggest that the mass should be one hundred times that of the proton. If experiments find the particle within the predicted mass range, the theory is confirmed. However, if they don't, theorists can always adjust the parameter of the theory so that the particle's mass is higher than masses that current experiments can probe. People may go on theorizing for a very, very long time without any guidance from experiments, a point made forcibly clear a while ago by physicist and writer David Lindley.[6] How would they know when to stop? Some may argue that, in the final analysis, what matters is the science we discover along the way. To a point, I agree with that. Think of Kepler, and of all that he achieved while searching for the mythic harmony of the world: nothing less than the three laws of planetary motion, verifiable statements about real phenomena. However, if achievements are minimal after a reasonable amount of time, we may begin to wonder about the wisdom of the quest. Analogies with sailors and explorers of the past are inspiring but flawed. If there were no Americas to be discovered, soon enough the Spanish and Portuguese ships would circle the globe and come home. *They set out to explore knowing full well that the globe was finite.*[7] This is very different from setting off on a mission without knowing if it even has a destination. Yes, ships must be sent off to explore the unknown, even if the costs are high. But it should not be forgotten that the notion of a Theory of Everything is cultural and not based on scientific evidence.

At present, superstring theory is the only viable candidate for a unified description that includes gravity and the other three known interactions. As such, it has received both enthusiastic support and criticism. Although I would argue that equating superstring theory with the Final Theory is akin to religious dogma and thus not science (since no unified theory can be proven to be final), I believe people should pursue it as long as it makes sense for them to do so. We don't want to give up on a promising idea, just as we don't want to keep on living a fantasy. Think of Albert Michelson, who in spite of his own findings died convinced that the luminiferous aether existed decades after it had been disproved. I do not intend to engage here in a point-by-point criticism of the good and bad of string theories. I think of them as a work in progress, as I am sure my competent colleagues working on them do. For all their promise, the theories face enormous conceptual challenges, the most urgent being the choice of a solution (a vacuum) corresponding to the real world, that is, a solution describing the particles and forces of the Standard Model in an expanding universe compatible with the cosmological Big Bang framework.

Two recent books have criticized string theory for technical and even sociological reasons: Lee Smolin's *The Problem with Physics* and Peter Woit's *Not Even Wrong*. As one can imagine, the books awakened much interest in the worthiness of the theory as a way toward unification. Responses from notable string theorists were of course highly critical of the books and their writers. Some were even offensive. I find this sort of dueling pointless. People should be free to research whatever they want, although they should also reflect responsibly on whether their goals are realistic. Einstein spent the last decades of his life searching for a unified theory of gravity and electromagnetism. Many physicists criticize him for that, saying he wasted his time. The criticism is not of Einstein's belief in a final theory, a belief many of the critics share, but for him having limited his search to only classical gravity and electromagnetism, leaving out the strong and weak nuclear forces. In other words, Einstein's critics believe that a final theory is achievable once we include the four known interactions. The number of fundamental forces changed, but the belief remains the same. I find it healthy and urgent to question the reasons behind the quest for a final theory and what they tell us of the way we think about the world and our place in it. Believers or not, we must all think critically about the wisdom of this search for Oneness in science and about what it has achieved in the past century.

To start with, I question the need for the tantalizing prize at the end. Do we need to believe in the Final Truth in order to pursue Nature's deep secrets? If we do, is it telling us something about Nature or about ourselves? Does the Universe need to be "beautiful" in order to be worth comprehending? Why insist on relating Oneness with beauty? Isn't it time to celebrate a different kind of beauty, one inspired by the imperfections of Nature? Quoting Wilczek: "Faith in the possibility of unification drives us into a state of denial. . . . Appearances—or rather, our interpretation of them—must be deceptive." Now, Wilczek goes on to argue that modern physics *is* pointing toward unification, that the hints are there and we are not delusional. We will get to look at such hints in due course. Even if my great admiration for scientists such as Weinberg and Wilczek makes me say this with some trepidation, I don't find these hints as convincing as they and many others do. The word *hope* appears quite often in their books. "I hope that string theory really will provide a basis for a final theory," wrote Weinberg.* "What shoes will unification including gravity drop? Any that we can hope to hear?" wrote Wilczek, more cautiously.† I find that there is indeed some delusion going on. The hints of a final theory—consisting mostly of the gaps in our present knowledge that many expect will be filled with the unification of all forces—are not as obvious or even as circumstantial as it is widely believed. I say this with a measure of sorrow. When I pursued unification I would hold on to the smallest scrap of evidence as enough to move on. But as time passed, each bit of exciting news that we were closer than ever to a believable string theory or to grand unification was either proven wrong or grossly inflated. Even though I knew well that science often advances through torturous paths, I started to question the

* As we can see from Weinberg's epigraph at the front of the book, things may be changing.

† I should stress that in *Lightness of Being*, Wilczek took great pains to separate fact from speculation. However, as he wrote to me late in 2008, "I'll be very disappointed in Mother Nature if she's been leading us on with a tease. We'll see."

whole unification enterprise. To make things worse, attempts to rescue failed theories felt forced and even more detached from physical reality. The whole thing started to feel more like faith than science.

The situation echoes, with great irony, the "God of the gaps" argument of the science and religion wars, which claims that God begins where science stops. As science progresses and we learn more about Nature, God, to His humiliation, gets squeezed into an ever-shrinking gap. Believers are convinced the gap will never completely close. Skeptics are convinced it will. The equivalent statement for unification, the "unification of the gaps," would go like this: unification begins where our current theories stop. What we don't know, unification will explain. As science advances and we learn more about Nature and its violation of symmetries, unification, to its humiliation, gets squeezed into an ever-shrinking gap. Theories are hastily revised, parameters are shifted, the whole mission of unification gets redefined. The discovery of dark energy is an apt example. Before 1998, theories of unification had as their express goal the cancellation of quantum fluctuations of the vacuum so as to neutralize their cosmological effect: the energy in these fluctuations would cause the Universe to expand at an accelerated rate, as did Einstein's cosmological constant (see Part II). After the discovery of the cosmic acceleration in 1998, a finite vacuum energy—a finite cosmological constant accelerating the Universe or other form of "dark energy"—seems unavoidable. Suddenly, people are trying to justify its value with unification arguments. Many even invoke the *anthropic principle*—that the Universe is the way it is so that life could flourish in it—to justify the observed value of the dark energy. What a reversal!

Wouldn't it be best to simply admit that our view of the world is a permanent work in progress? That there is no Final Truth to be discovered for the simple reason that we will never have the totality of information to determine if we have arrived at the Final Truth? As I mentioned above, many Unifiers often say that Einstein could never have succeeded in his unification quest because he left out quantum mechanics and the strong and weak nuclear forces, which were not well understood until after his death. They then go on to argue that now things are different, that now we know better. Well, how can we be so sure? How do we know that there aren't other forces out there waiting to be discovered, maybe a deeper layer of particles and interactions? How can we be sure that there won't always be some new piece of knowledge lurking in the shadows, beyond the reach of our detectors, rendering our unification efforts forever incomplete? Theories may guide us, but no more than that. Only experiments can decide what is real.*

What we know of the world is surely not all there is to know. Any assertion to the contrary displays only human arrogance. Hence unifying all there is—even at the level of fundamental physics—is fated to fail. The belief that human thought can grasp a final truth is a faith-based fallacy that feeds on our need for being more than human, for being omniscient as gods, a fallacy that reverts back to our fear of loss and our all-too-many limitations. Final unification, even within the limited realm of particle physics, is impossible. The best that we can do is collect what we discover of the world in the most coherent possible way. Once we accept this we can admire Nature for its imperfect beauty, ever creative and ever surprising. Once we accept this we can look at Nature with human eyes and not through the lenses of omniscient-god wannabes. To paraphrase Socrates, the more we know, the more we should be humbled.

I propose that we focus on the imperfections of Nature rather than on a search for ultimate perfection. This new approach, as we shall see later on, has consequences beyond science; it forces us to look at the world for what it is, not for what we want it to be. Symmetries are great tools but they're not the final law. Nature is beautiful for being imperfect. Behind every imperfection there is a mechanism for generating structure and complex behavior. Imperfection and imbalance are the seeds of becoming. Perfect Nature would be stale and formless, existing only in a Platonic realm, detached from reality. Again, dark energy is a perfect example. It is "ugly" and unexpected; its value contradicts common sense. Yet it makes the cosmos flat enough to generate galaxies and, eventually, life itself.

In order to be more concrete about the role of asymmetry in the world of the very small, we'll take a critical look at unification's achievements and examine some of the many questions it leaves unanswered.

* And even here we must be careful; experimenters decide the quantities that will be measured and how to calibrate the filters that throw away presumably "undesirable" data.

32

SYMMETRIES AND ASYMMETRIES OF MATTER

The first unification was that of electricity and magnetism. Solutions to Maxwell's equations in the absence of electric charges and magnets display a beautiful symmetry between electricity and magnetism: the two fields bootstrap each other as they propagate along empty space at the speed of light. But once sources are present, the symmetry is imperfect; magnetic monopoles, the analogues of static, individual electric charges, do not exist. If they do, they have certainly eluded our best efforts to find them.[8]

Moving on to particle physics, we encountered the matter-antimatter asymmetry. In the language of internal and external symmetries, there exists an internal symmetry operation (in practice, a mathematical operation) that changes a particle of matter into one of antimatter.* The operation is called *charge conjugation,* and is represented by the capital letter *C*. The observed matter-antimatter asymmetry implies that Nature does not display charge-conjugation symmetry: in some cases, particles and their antiparticles cannot be turned into one another. Specifically, C-symmetry is violated in the weak interactions. The culprits are the neutrinos, the strangest of all known particles. It's time to tell their story.

Right before World War I, a strange experimental result added to the many nightmares stealing the sleep of quantum physicists. James

* A mathematical operation can be very simple, as when we add 1 to another number, or when we turn a cube in our hands. Although the operations used in particle physics are more complex than these, in the final analysis they are doing things to numbers or to geometrical objects.

Chadwick, who in 1932 discovered the neutron, was investigating beta decay, the emission of electrons from radioactive nuclei. Nuclei with too many neutrons may improve their stability by transforming a neutron into a proton. From charge invariance, the proton's positive charge must be balanced by a negative charge. Out goes the electron to keep charge conservation in check. The shock came when Chadwick tested the sacred conservation of energy. The beta-decay electrons should have a fixed amount of energy, the mass difference between the two nuclei times c^2.* They didn't. Their energies varied broadly. Some were emitted with high speeds, others with low speeds. No one understood why. As late as 1929, Niels Bohr wrote: "We have no argument for upholding [the law of energy conservation] in the case of beta ray disintegrations. The features of atomic stability responsible for the existence and properties of atomic nuclei may force us to renounce the very idea of energy balance."

Take note: the great Niels Bohr was ready to give up on energy conservation! The situation was clearly getting desperate. Rutherford was more cautious and decided to wait and see. A hopeful Dirac declared: "I should prefer to keep rigorous conservation of energy at all costs." So would everyone else. But how?

Late in 1930, Wolfgang Pauli came up with a crazy idea. He wrote in his journal: "I have done something very bad today by proposing a particle that cannot be detected; it is something no theorist should ever do." He sent a letter to colleagues gathered in Tübingen to discuss radioactivity: "Dear radioactive ladies and gentlemen. I have hit upon a desperate remedy. Namely, the possibility that there could exist in the nuclei electrically neutral particles that I wish to call neutrinos. . . ."[9] The neutrinos would have varying energies so that, when added to the energy of the emitted electron, the total would match the mass difference between the two nuclei, and everything would work according to plan. Energy conservation was saved.

In fact, *anti*neutrinos and not neutrinos are emitted in beta decay. The reason is another internal symmetry that is conserved in the

* Write the decay as nucleus-1 ⟶ nucleus-2 + electron. Using $E = mc^2$, conservation of energy dictates that the emitted electron's energy *must* equal the difference in masses of the two nuclei (times the square of the speed of light), if that was all there was to it.

weak interactions: lepton number.[10] We can think of lepton number as a kind of charge, like the electric charge. Every lepton (such as the electron) carries one positive unit of lepton number; every antilepton (such as the positron) one negative unit. If lepton number is conserved in beta decay, the lepton number of the electron (+1) must be canceled by the lepton number of the antineutrino (−1). And so beta decay is described as

$$\text{neutron} \rightarrow \text{proton} + \text{electron} + \text{antineutrino}$$

Since neutrons and protons are hadrons, their lepton numbers are zero. The math works out nicely: on the left-hand side, the total lepton number is zero (the neutron); on the right-hand side, it is also zero.

Neutrinos, interacting only via the weak force, are extremely hard to detect. They are produced profusely in the interior of the Sun, as hydrogen fuses into helium. Every second, trillions of these solar neutrinos pass through your body without you knowing it. There is much more to our connection with the Sun than daylight and warmth.

In spite of their ghostliness, neutrinos can be detected. And in 1956 they finally were. Twenty-six years passed between Pauli's prediction and the detection of neutrinos. This lag between theory and experiment is often used as an illustration of why sometimes we just have to wait a while to make profound discoveries. Technology usually trails theories, at least in high-energy physics and cosmology. Ideas are cheaper than machines. The Higgs particle, for example, was proposed more than forty years ago and still hasn't been detected. Hopefully, the LHC will discover it, or something like it, as we discussed in Part II. However, I would be cautious with arguments using lag times between theory and experiment as comfort for the current dearth of data on unification. We can't learn particle physics from history. As we shall see, if neutrinos prove anything, it is how asymmetric Nature is. They are the torchbearers of the imperfect cosmos. To understand why, we must introduce a different kind of spatial symmetry called *parity*.

A parity operation, represented by the capital letter *P*, turns an object into its mirror image. You can't effect this transformation through translations and rotations. Our faces are approximately parity-invariant (neglecting little blemishes and moles) but our bodies are not: your mirror image has the heart on the right-hand side.[11]

Particles have spin: they rotate about themselves like tops or the Earth. However, particles are no ordinary tops. Being quantum objects, their spin is quantized: a particle can only rotate in a small number of ways. In contrast, a top can spin at arbitrary rotations per minute (rpm). It's a bit like old vinyl records, playable only at 33⅓, 45, or 78 rpm.* All particles of matter, that is, all quarks and leptons, can rotate in only two ways. We say they have spin ½, the smallest amount of rotation possible, the quantum of rotation.[12] To simplify things, imagine a particle rotating at the same rate with respect to the vertical direction either to the left or to the right. One rotational sense is actually the mirror image of the other. You can experiment in front of a mirror with a ball or a screwdriver to confirm that. The parity symmetry operation, when applied to a spinning particle, can flip the direction of its spin.

Beta decay, which brought us neutrinos, had another ace up its sleeve: it can be used to show that neutrinos are not symmetric under parity. Nature is not mirror-symmetric; it has a preferred spatial orientation. It's as if you had a top that would only spin counterclockwise! In 1956, two Chinese-American physicists, T. D. Lee and C. N. Yang, predicted that the weak interactions would violate parity invariance. To the dismay of many, in a few months T. T. Wu and her team confirmed their prediction. Neutrinos only interact with matter in their *left-handed form:* if you imagine that they are moving upward, they only spin from east to west. In turn, antineutrinos only appear in right-handed form: they spin from west to east. Nature has a very obvious preference for handedness.

Right-handed neutrinos may exist, but they have never been detected. This means that either they interact incredibly weakly with ordinary matter or they are very heavy. (Or they simply don't exist.) Either way, they are obviously quite distinct from their ubiquitous left-handed cousins. So much for the high symmetry of particle physics. But wait! There is more.

Let's put C and P together. (That is, charge conjugation and par-

* A word of caution: *rotation* is a classical word. It's tempting, and sometimes useful, to visualize particles as little spinning balls, but that's not what they are. Having said that, I ask you to please go on visualizing them that way.

ity.) Applying the C operation on a left-handed antineutrino, we should get a left-handed antineutrino. The problem is, there are no left-handed antineutrinos in Nature. This is why the weak interactions, the only interactions neutrinos feel (apart from gravity), violate charge conjugation symmetry. Now let's go one step further. If we apply *both* C and P to a left-handed neutrino we should get a right-handed antineutrino: the C flips neutrino into antineutrino and the P flips left-handed into right-handed. And yes, antineutrinos are right-handed! We seem to be in luck. The weak interactions violate C and P separately but apparently satisfy the combined CP symmetry operation. In practice, this means that reactions involving left-handed particles should occur at the same rate as reactions involving right-handed antiparticles. Everyone was relieved. There was hope that Nature was CP-symmetric in all known interactions. Beauty was back.

The excitement didn't last long. In 1964, James Cronin and Val Fitch discovered a small violation of the combined CP symmetry in the decays of a particle called *neutral kaon*, represented as K^0. Essentially, K^0 and its antiparticles don't decay at the same rate as a CP-symmetric theory predicts they should. The physics community was shocked. Beauty was gone. Again.

CP violation has an even deeper and more mysterious implication: particles also pick a preferred direction of time. The asymmetry of time, the trademark of an expanding Universe, happens also at the microscopic level! This is huge. We need a new paragraph to deal with it.

That time moves forward should be pretty obvious to everyone. We make eggs turn into omelets and not vice versa. A sugar cube stirred into coffee doesn't spontaneously reassemble back into a cube. Plants don't go from flower to seed. We don't grow any younger. If we made movies of someone cooking or a plant growing and played them backward it would be clear that the time direction was reversed. For simpler systems, however, the distinction is not obvious. A pendulum oscillates left to right and right to left. Looking only at the swinging motion we wouldn't know what the proper time direction was. The same is true of two billiard balls colliding or a photon hitting an electron. Systems like these are called *time-reversal invariant*: they do not have an obvious arrow of time. We can create a symmetry operation, *time-reversal,* that reverses the direction of time of a system. It's represented by the capital letter *T*. If a ball is moving from left to right, after applying T it will move from right to left. In Part II, we saw that the expansion of the Universe breaks time-reversal invariance at the astronomical scale: there is a cosmic direction of time linking the origin of galaxies and ultimately our own to the Universe as a whole. In the subatomic world, things were supposed to be different; they were supposed to be as symmetric as possible. Well, they aren't. The Standard Model of particle physics must incorporate these asymmetries.

There is another way to see the connection between CP violation and the arrow of time. The theories of particle physics *must* obey the combined triple symmetry CPT. (Yes, apply all three of them in sequence: particle to antiparticle, left-handed to right-handed, and motion forward in time to backward in time.) Abandoning this symmetry would mean that Einstein's special theory of relativity, the backbone of all our theories describing the interactions of matter particles, is wrong or at least severely incomplete. Fortunately, so far no CPT violation has ever been observed. That being the case, and since CP is violated, then T must also be violated so that the product of the two remains invariant. Like multiplying −1 with −1 to get 1. So, as long as CPT holds, violation of CP implies a specific arrow of time at the microscopic level.

Nowadays there are examples of CP violation for another particle family, the B mesons. Nature not only picks a handedness but also a sense of time. An open question is why this only happens with the weak interactions. Many expect the strong interactions to violate CP also, but there is no experimental evidence that it does. Quite to the contrary, the evidence is that it doesn't. Several explanations have been proposed for the lack of CP violation in the strong interactions. The most popular predicts the existence of a light particle called an *axion*, but despite much searching, axions haven't been found. That the strong and the electromagnetic interactions are CP-conserving makes them substantially different from the weak force. Attempts to unify the three must find ways to overcome this, not a trivial task.

33

THE ORIGIN OF MATTER IN THE UNIVERSE

From my perspective, CP violation is a precious gift. It gives us a chance to understand why there is more matter than antimatter in the Universe, starting from an initial situation with equal, or near equal, amounts of each: the asymmetry developed in time, as the Universe evolved. The alternative would be a Universe that accidentally emerged from the primordial quantum soup with an excess of matter over antimatter in the correct range. In this case, the "one-in-a-billion" excess of matter over antimatter particles would be due to some unexplainable initial condition. Given the chance, physicists like to understand the mechanisms behind natural phenomena. The matter-antimatter excess is no exception.

The first to propose a connection between the matter excess and CP violation was the great Russian physicist and peace activist Andrey Sakharov. In 1967, just three years after Cronin and Fitch discovered CP violation, Sakharov wrote a prescient paper in which he outlined the three conditions for a matter excess to evolve in the early Universe. In order to generate a matter surplus, the interactions among the particles must make more quarks than antiquarks and the surplus must be preserved as the Universe expands. In more detail:

1. *There must be baryon-number violation.* The same way that the electron and the other leptons carry lepton number, baryons carry baryon number. So, a proton and a neutron each have baryon number +1. Their antiparticles have baryon number −1. Clearly, in order for interactions among particles to produce either more baryons or antibaryons they must violate baryon

number. The net baryon number of the particles before and after they interact should not be the same. If it grows there will be an excess of baryons. If it decreases, the excess is of antibaryons.[13]

2. *There must be a violation of charge conjugation and of CP.* Just making either more baryons or antibaryons is not enough. We need a bias, a preference toward more baryons. This is achieved using C and CP violation. The amount of violation is critical in determining the final matter-antimatter asymmetry.
3. *There must be thermal nonequilibrium conditions.* Recall that in thermal equilibrium everything remains the same on average. So, if the early Universe was in thermal equilibrium when more baryons were being made, antibaryons would also end up being made and the excess would be erased. In order to preserve the baryonic excess made with conditions 1 and 2, the Universe must have remained out of thermal equilibrium for a while.

How does the Universe get out of thermal equilibrium? Recall the discussion in Part II of the amazingly homogeneous microwave background temperature and the related horizon problem in the Big Bang model: thermal equilibrium is always established through collisons between particles. In cosmology, thermal equilibrium is usually defined by comparing two (or sometimes more) time scales: in this case, the time scale of the cosmic expansion (that is, how fast the Universe is expanding) and the time scales in which the particles interact with each other. If the Universe expands faster than the particles can interact with each other, they won't be able to exchange information and keep at the same temperature: the expansion pushes them apart too fast. The reactions are occurring out of equilibrium. Whereas when particles interact faster than the expansion, they remain in thermal equilibrium.

Could all three conditions be satisfied together sometime during the cosmic infancy? The answer is possibly yes. The first models of *baryogenesis*, which attempted to explain the "genesis of baryons," applied the three Sakharov conditions in the context of Grand Unified Theories (GUTs), which were proposed in the mid-1970s to unify the strong, weak, and electromagnetic forces. Recall that the strong interactions explain how quarks interact via exchanges of gluons, while the weak interactions explain radioactive decay via the existence of C and CP-violating processes involving the three weak gauge bosons. To unify the forces is to remove the distinctions between quarks and leptons. In other words, in a grand unified world, quarks can change into leptons and vice versa. As a consequence, the proton is no longer stable: diamonds may not be forever. The original GUT model proposed by Sheldon Glashow and Howard Georgi in 1974 predicted that the proton would decay in about 10^{30} years, a time a trillion billion times longer than the age of the Universe. "Nonsense!" protests the reader. "That's the same as saying that the proton is stable." Well, not quite. All you have to do is gather enough protons in a large enough volume and see if some decay within a reasonable amount of time.*

Experimental physicists, eager to confirm that GUTs were correct, collected huge amounts of water (over ten thousand tons) in underground tanks covered with sensors that could detect the elusive proton decay. Several searches went on across the globe. But the proton didn't decay. Rocky Kolb, who was my postdoctoral advisor at Fermilab and one of the leading cosmologists working on GUT baryogenesis in the early days, once told me of the "devastating feeling when we found out that the proton had not decayed. We were so sure it would happen. . . ." Theorists rushed to revamp their models, increasing the prediction for the proton lifetime by a few orders of magnitude to make them consistent with experimental limits. As the detectors grew, the proton still refused to decay. All simpler GUT models failed to past the test. Ockham's razor does not seem to be at work here.

Some GUT models can still elude the current experimental limits. The price, however, is to either make them more contrived or invoke a new symmetry of Nature called *supersymmetry*. This symmetry, as the name already says, is truly super: it makes particles of matter into particles of force. Proposed in the early 1970s, it would be the grandest of all symmetries, connecting every possible kind of particle. Not surprisingly, Unifiers are very keen on supersymmetry, affectionately referred to as SUSY.[14]

* Recall Avogadro's number: in 12 grams of carbon there are 6×10^{23} atoms, already about a trillion trillion protons or so. In a volume with 10^{30} protons we should expect at least one decay a year.

SUSY theories make impressive predictions about the matter content of the Universe. Specifically, every particle of matter should have a supersymmetric partner: a photon has a "photino," a gluon a "gluino," a quark a "squark," etc. You get the idea. If SUSY is confirmed, the number of elementary particles automatically doubles, in similar fashion to what happened when antimatter was discovered. Since, unlike antimatter, none of these supersymmetric particles has been observed, they should all be extremely heavy or unstable. If they are very heavy, they can't be made in current accelerators. If they are highly unstable, they may decay before being detected, although the decay pattern may tell us something of their properties. If this were all, the situation would be quite hopeless. Untestable ideas don't make much sense in science. Fortunately, many models predict that the lightest of SUSY particles is stable. This lightweight SUSY particle is second only to the Higgs on the most-wanted list of elusive particles of Nature. If found, it would prove that SUSY is real, promoting it into a plausible solution to many of the puzzles now haunting particle physics and cosmology. The particle is a leading candidate for dark matter, for example. So far, and in spite of much effort from dozens of experiments across the world, it has escaped detection, forcing theorists to push its mass to higher values and to limit its range of interaction.

Many believe the LHC is likely to find it. All bets are off, really, since SUSY may be just a clever invention of symmetry-hungry theorists. The fact that the particle has so far eluded detection doesn't bode well.* To make things worse, results from the giant Super-Kamiokande detector in Japan and the Soudan 2 detector in the United States have ruled out supersymmetric GUT models, at least the simpler ones, based again on the proton lifetime. If SUSY is a symmetry of Nature, it is very well hidden. We know that it must be "broken," that is, it can't be a symmetry at the energies that current experiments probe; otherwise we would have seen it. The details of how this symmetry breaking happens, which are unknown, are directly tied to the masses of the hypothetical superpartners, the supersymmetric partners of ordinary matter particles. This allows

* Of course, within a few years after the publication of this book, the lightest SUSY particle may be found. I remain very skeptical.

theorists to push SUSY breaking to energies beyond the reach of experiments in the foreseeable future, a very unsettling possibility.

Given the difficulties with GUTs, we must search for other ways to make the matter excess in the early Universe. After all, we are living proof that it is real. Fortunately, there is another way. Instead of appealing to hypothetical grand unification, why not use the fact that the weak interactions already violate C and CP (Sakharov's condition 2) and try to generate the matter excess at lower energies, where the Standard Model describes the physics? Much more reasonable and concrete. From a cosmological perspective, this would delay the period of baryogenesis from one-trillionth-trillionth-trillionth of a second after the Bang—the time before which the unification of the strong, weak, and electromagnetic forces presumably held (about the same time that inflation took place; see Part II)—to one-trillionth of a second, still in the time before which the unification of weak and electromagnetic interactions held. The challenge is to satisfy conditions 1 and 3, that is, to find a way of violating baryon number in the Standard Model and to have departures from thermal equilibrium. In the next chapter we'll look at how this could work.

34 A UNIVERSE IN TRANSITION

"Baryogenesis at the electroweak phase transition" is a mouthful, but we have learned enough in this book to parse it. *Baryogenesis:* the production of more baryons than antibaryons, that is, of more matter than antimatter. *Electroweak:* the force resulting from the unification of the electromagnetic and the weak interactions, our next topic. *Phase transition:* processes where a change in external conditions leads to a qualitative change in a system, as when water turns to ice following a temperature drop. We will see that processes analogous to phase transitions also describe qualitative changes in the internal symmetries of particle physics.

The Standard Model embodies all that has been achieved so far on the quest toward unification. We mentioned before the Nobel Prize–winning work of Glashow, Salam, and Weinberg, which predicted the existence of the weak gauge bosons responsible for the weak interaction, the W^+, W^-, and Z^0. Their discovery in 1983 heralded a new era of particle physics, giving the Standard Model the credibility it now has. The theory went beyond the prediction of new particles, proposing a new way to think about mass. Recall the Higgs, the hypothetical field whose role is to give mass to all particles of matter and of force. The field is everywhere, omnipresent across the Universe. According to the Standard Model, there are two possible phases for particles to be in, determined by the Higgs field. If its value is zero, all particles are massless. If it's nonzero, all particles acquire a mass. The strength of a particle's interaction with the Higgs sets its mass: the stronger the interaction, the larger the mass. The photon is the only exception, now that we know that neutrinos have small masses. (We still don't know their values, though.)[15]

How did this dramatic switch in the properties of the particles (massless to massive) take place? The answer is found in phase transitions. We live in the ("frozen") phase where the Higgs is nonzero and the particles are massive. That's the low-energy phase. According to present estimates, at energies above roughly two to three hundred times the mass of a proton (times the speed of light squared) the Higgs field becomes a sort of transparent presence, invisible to all the particles. Since particle interactions with the Higgs sets their masses, all particles become massless. That's the high-energy ("liquid") phase.

Back to the water and ice analogy: note that water and ice have very different spatial (external) symmetries. While water is homogeneous, that is, on average it looks the same everywhere, ice is inhomogeneous: the frozen water molecules occupy very specific spatial locations. In fact, they form a beautiful hexagonal lattice, like a honeycomb structure. Oxygen atoms sit on each of the six vertices and the two hydrogen atoms along the two lines connecting the vertices with each other. This sixfold symmetry of ice crystal lattices determines the beautiful sixfold patterns of snowflakes, macroscopic manifestations of a microscopic symmetry. Even though crystals have a high degree of symmetry, liquid water is even more symmetric since it looks the same everywhere: the average probability of finding a water molecule is the same across the volume. Thus, as the temperature drops and water transitions from a liquid to a solid phase, it loses symmetry: the phase transition prompts a decrease in symmetry.

Something similar happens with the Higgs field and the electromagnetic and weak interactions. When the Higgs is transparent, the weak gauge bosons are massless just as the photon is: the weak interactions are then long-range and behave somewhat like electromagnetism. For this reason, we say that in the high-energy phase the two interactions are unified into the electroweak force. At low energies, the Higgs loses its transparency and interacts with all particles of matter and of force, giving them their masses. The only exception is the photon, which remains massless throughout. The weak gauge bosons become very heavy, forcing the weak interaction to work only at a very short range. As a result, it detaches from electromagnetism. As with water and ice, there is a loss of symmetry in going from the high-energy to the low-energy phase. Since in the high-energy phase the two forces behave similarly, the symmetry is higher than in the low-energy phase, the reality we live in, where the two forces are very different. The subtlety is that the symmetry is not spatial but internal, being related to the special charges that particles interacting via the weak and electromagnetic forces carry. This loss, or breaking, of symmetry is the signature of the *electroweak phase transition.*

In 1985, as I was finishing my Ph.D. in London, three Russian physicists, Vadim Kuzmin, Valery Rubakov, and Mikhail Shaposhnikov, published a bombshell of a paper.[16] Their very smart idea was to use the electroweak phase transition as the source of matter excess. To do that, they went back to cosmology, realizing that the Big Bang model predicts that the Universe is hot early on. Thus, just as ice melts when heated up, the Higgs field would also be heated up at early times. The effect of this heating is to force the Higgs toward its transparent, high-energy phase. Going back early enough, to times before one-trillionth of a second after the Bang, the Higgs was hot enough to become transparent: the broken symmetry that prevents the weak interactions from becoming long-range in our reality was restored! We know that ice melts at 32 degrees Fahrenheit. The Higgs becomes transparent at energies approximately two hundred times the mass of the proton (times the square of the speed of light). The consequence is inescapable: the early Universe went through a phase transition.

The electroweak phase transition helps solve the matter excess problem in two ways. First, it provides a way for baryon number to be violated early on in the life of the Universe (Sakharov's condition 1), something that is highly suppressed at low energies. Imagine that you work in a Kafkaesque office complex, an infinite chain of cubicle after cubicle separated by very thick walls, each ten feet tall. Just as in a usual office building where each office has an assigned number, each cubicle has an assigned baryon number that differs from its two nearest neighbors by three units. To the right, baryon number grows by three units; to the left, it decreases. Each cubicle thus correspond to a different "world," defined by its baryon number. At ordinary energies and temperatures, the only way to move from one cubicle to the next (and gain or lose three units of baryon number depending which way you go) is boring a hole through the wall. Since there aren't any digging tools in your cubicle, it would take you a very long time to dig a tunnel using your nails, longer than an average human life. Not impossible, but highly improbable (and painful). As you start to resign yourself to being stuck in the same cubicle forever, you remember that every cubicle has a stool mounted on a spring. Excited, you discover that a temperature-sensitive latch locks the spring. At a high enough temperature, the latch will release the lock and the spring will propel the stool upward with great force. The rest is easy. You stack pages upon pages of wrong calculations under the latch and start a huge bonfire. As you wait for the fire to spread, you jump onto the stool. After a while the latch releases the spring and off you go, flying into the next cubicle.

The Kafkaesque cubicles capture the essence of the Kuzmin-Rubakov-Shaposhnikov mechanism. At low temperatures, baryon number can be violated in the Standard Model but only at a very tiny rate. Although the walls separating "worlds" with different baryon numbers (the cubicles) can in principle be tunneled through via quantum processes, in practice they aren't. That's a good thing, since otherwise the proton could decay and we wouldn't be here to think about baryogenesis. At high temperatures, however, baryon-number-violating processes are less suppressed: "worlds" with different baryon numbers can be accessed without much difficulty. The hot early Universe naturally offered the high temperatures needed to promote these jumps. Sakharov's first condition is satisfied.

If one now adds the C and CP violation from the weak interactions, the picture gets even better. Their net effect is to tilt the array of cubicles in one direction, as if they were built on a hill. This way, a "downhill" direction for baryon number violation is favored and more matter than antimatter is produced. Sakharov's second condition is satisfied.

What about the nonequilibrium conditions needed to ensure that the net matter excess is not erased? Those come from the phase transition itself. Think again of water turning to ice. Initially, the water is nice and homogeneous and there is no sign of ice crystals. As the temperature drops, we start to notice small little bundles of frozen water. The tiny ice crystals are the seeds, or nucleating sites, of the transition. In fact, their formation is often prompted by the presence of an impurity. This is what happens when it rains or snows. Water vapor from cooled air condenses around a dust particle. If the condensation happens close to the ground, we see it as dew droplets. If high up, we see it as clouds. When the temperature drops low enough, a droplet may freeze, prompting its neighbors to do the same. As ice is colder than water, the freezing releases heat. This is a typical out-of-equilibrium process. Equilibrium is restored only when the whole system converts to the new phase. As you can see in your freezer, once all the water converts to ice, not much else happens to it.

Back to the electroweak phase transition: there are two phases, the high-temperature, symmetric phase and the low-temperature, asymmetric phase. Particles in the symmetric phase are massless, while they are massive in the asymmetric phase. Recall that in the high-temperature phase, baryon number can be violated and a net excess of matter can be generated: walls separating cubicles can be jumped over. Start in the early hot Universe, when the Higgs was still transparent. Particle reactions could jump from cubicle to cubicle and baryon number was violated. The C and CP violations ensured that more baryons were being made. While this was happening, the Universe kept expanding and cooling down. Eventually it cooled below the Higgs condensation threshold and the symmetry was broken: jumps to neighboring cubicles were suppressed and the transition to the asymmetric phase started. How did the transition happen in different locations? Is there an analogue to the condensation of tiny ice crystals? The answer depends on how heavy the Higgs is, something we still don't know.

I got into baryogenesis research somewhat late, around 1990, when I was a postdoctoral fellow at the Institute for Theoretical Physics in Santa Barbara. (Before that, I was still pursuing research on the cosmological consequences of superstring theories.) At the time, we all thought that the Higgs could be quite light, lighter than the weak gauge bosons, with a mass around forty or fifty times that of the proton. Were that the case, the transition from symmetry to asymmetry would be of the "first-order" (or discontinuous) type: in a sea of transparent Higgs and baryon-number-violating processes—the realm of electroweak symmetry and cubicle jumping—a small bubble of the asymmetric phase would pop up. (The equivalent of the ice crystal seeds appearing in water.) That bubble, if large enough, would grow and meet other similar bubbles. Eventually the expanding bubbles would fill up most of the volume of the Universe and we would be in the heavy Higgs phase: all particles but the photon would have masses and baryon number would be conserved. The phase transition would be complete.

The mechanism to generate the matter excess takes advantage of these expanding bubbles. Picture what was going on outside and inside the bubbles. Outside, in the symmetric ("liquid") phase, baryon number was being violated and an excess of matter particles was being created. Inside the bubbles, in the asymmetric, low-energy world, no such processes were allowed. Here comes the trick: the bubble walls were not completely opaque. Like sperm penetrating an egg, matter and antimatter particles from the outside could occasionally come through. (I'm hoping this analogy won't be easily forgotten.) Since there was an excess of matter particles outside (as if there were more male—matter—than female—antimatter—sperms, as seems to be the case in my family), these would come through the bubble wall in larger numbers and produce the observed matter excess. Voilà! We have a beautiful physical mechanism to explain why there is more matter than antimatter in the cosmos!

In 1993, I co-authored a paper with Rudnei Ramos, then a postdoctoral fellow from Brazil visiting my group at Dartmouth, which predicted that this scenario based on bubble nucleation would break down if the Higgs were more than seventy times as massive as a proton. Soon after, Shaposhnikov (one of the three Russians that proposed the electroweak baryogenesis scenario we discussed) and collaborators confirmed our results (and went well beyond what we did) using large-scale computer simulations to study the details of the electroweak phase transition. I was devastated when, a few years later, the Higgs was shown to be at least 105 times heavier than the proton. The simple bubble nucleation scenario was ruled out. I insisted, showing that even for a Higgs this heavy, a variant scenario called weak first-order phase transition could work. But the Higgs mass kept creeping up, making life very difficult. The conclusion was unsettling: we can't use the Standard Model in its present form to obtain the needed matter excess during the electroweak phase transition. Of course, someone can always come up with a bright new idea—and there are many scenarios in the literature that extend the Standard Model—that would do the job. At the moment, the most popular alternative is, not surprisingly, to invoke SUSY. When supersymmetry is incorporated into the Standard Model, the matter excess production can be enhanced. However, even in this case, the simpler models seem to fail. I tried a few of them with my friend and colleague Mark Trodden in 2001. The truth is, we still don't know how to explain the matter excess in the Universe, although most of us feel that we are inching toward a proper solution. The search, though, has been exhilarating.

35 UNIFICATION: A CRITIQUE

From Thales to Kepler to Einstein to superstrings, the quest for the Final Truth has inspired some of the greatest minds in history. Although superstring theory is a work in progress, and may be for hundreds of years, the quest has so far failed. True, some partial unifications have been achieved. We mentioned how electricity and magnetism behave as a single wave propagating across space at the speed of light. We also mentioned how the absence of magnetic monopoles spoils the perfection of this unification, although we can still treat electromagnetism as one interaction. We have seen how the weak interactions violate a series of internal symmetries: charge conjugation, parity, and even the combination of the two. The consequences of these violations are deeply related to our existence: they set the arrow of time at the microscopic level, providing a viable mechanism to generate the excess of matter over antimatter. Without these asymmetries, the universe would be filled with a soup of radiation and a few sparse particles: no atoms, no stars, no people. The message from modern particle physics and cosmology is clear: we are the products of imperfections in Nature. Even if, to the accuracy of our measurements, other symmetries are indeed respected—energy conservation, electric charge—we still must face the fact that many aren't, or that they hold only as approximations. Asymmetries are the links to our origins.

Grand Unified Theories make two important predictions: the proton should be unstable and decay, and there should be new types of magnetic monopoles, heavy cousins of the simpler electromagnetic ones. After decades of searches in laboratories across the globe, no proton has decayed and no magnetic monopole has been detected.

We can always say that one day they will, that our present models are too simplistic and our detectors not sensitive enough. In the case of GUT monopoles, cosmic inflation can get rid of them, essentially keeping only one or a few within our observable Universe. Still, as time passes and increasingly accurate experiments force the models into an ever-shrinking gap, it's hard not to feel that something may be very wrong with the whole picture.

Then there is electroweak unification, our only model where two forces do behave similarly above a certain energy threshold and experimental data confirms some of its main predictions. There is no question that the theory is a triumph of modern physics; we already have sung its praises. But a careful look at the details shows that the electroweak unification is not a *true* unification. At least not in the sense of grand unification, which predicts that all forces become one. The electroweak theory never truly gets rid of the difference between electromagnetism and the weak interactions. The neutral-force-carrying particles we identify with the massless photon and with the heavy Z^0 at low energies are mixtures of the gauge bosons of the high-energy theory.[17] Furthermore, the left-handedness of the neutrino makes the theory lopsided: to be consistent with experiments, right-handed particles are described very differently from their left-handed counterparts.

The Standard Model is a wonderful achievement and should be celebrated as such. But it is also a compelling illustration of how Nature acts through imperfections and approximate symmetries. We have built a description, an incomplete narrative that is consistent with our current experiments. There are many holes in the Standard Model, features that remain unexplained. The neutrino masses are one example, the elusive Higgs field another; the reason why the mass of an electron is so different from that of a proton while their electric charges are the same (and opposite) is another. Many hope that the solutions to these problems will point toward a deeper theory, more in tandem with true unification. Supersymmetry is the most popular of such solutions, and we must wait to see if it will pan out. If the Large Hadron Collider or some dark-matter search experiment confirms SUSY within a few years, a major revolution will be launched in our understanding of the Universe. The micro and the macro will be woven even closer together. The possibility of a Grand

Unified Theory will be ever more tangible. Unifiers across the world will justly celebrate. Yet such unification, even if possible, will never be a true final unification. As I have been arguing, the Final Truth is a construction of the human mind, a monotheistic myth that has inspired Thales, Kepler, Einstein, and so many others to this day, with little support in physical reality. The alternative—unthinkable and even offensive to many, I am sure—is that we will never arrive at such theories, that such unification simply doesn't exist. All we can do is improve upon our narrative and describe Nature with our incomplete theories, probing deeper and deeper into its mysteries. Fundamental physics is still exciting without a Holy Grail.

In my early career working as a Unifier, Einstein was my main inspiration, but not the only one. Other pioneers of physics were also searchers. Heisenberg, Pauli, Schrödinger . . . How could they all be wrong after being right about so many things? I published some sixty papers on related topics, went to countless conferences across the world, gave hundreds of talks, worked for a decade on the rarefied strata of higher-dimensional theories and unification. Together with my Ph.D. advisor, John G. Taylor, I even wrote one of the first papers on how superstrings could explain the Big Bang, back in 1985. In spite of all this activity, in the early 1990s I started to feel estranged from the main current. I worried that many of the ideas related to unification were so far removed from experiments that they could never be directly tested. If that were true, how would we ever know if they make any sense? Is indirect evidence enough for physics? There is a huge gamble in dedicating an entire scientific career to ideas that may never be directly validated. But the prize is so high, the idea so alluring, that many choose to go this way. Should I?

Though I fought it initially, the notion that unification was just a fantasy began to take hold. Then, in 2002, my wife and I built a house in the middle of the New Hampshire woods, about eighteen miles south of Dartmouth College: no nearby neighbors, only solemn Mount Ascutney in the distance and the mighty Connecticut flowing below, impervious to our questions. Nature peered in through the vast glass windowpanes, impossible to ignore. For the first time in my life, I looked at the world with my eyes wide open, without a preconceived theory to lean on. I saw that trees never fork perfectly, that clouds are never perfect spheres, and that stars are scattered in the skies without any apparent pattern. I realized that we were imposing order on Nature, an order we longed for ourselves. There are natural laws, and they reflect patterns of organized behavior. But are these laws blueprints of physical reality? Or are they logical descriptions that *we* create to represent it? What have we learned in the recent past about our origins? That the Universe is expanding at an accelerated rate, that time had a beginning, that the reason we exist at all can be traced back to a fundamental imbalance in the way the particles of matter interact with each other, that only through random genetic mutations can life thrive and adapt. We have learned that without imperfections neither atoms, nor galaxies, nor people would exist. Yet in the face of all this mounting evidence, many of my colleagues continued (and continue) to believe in the abstract perfection of a Final Truth, smitten by the Ionian Enchantment.

In the winter of the same year that we moved to the house in the woods, I went for a walk with my daughter under the full moon. As we were crossing an open snowfield, she picked a handful of snowflakes and made them glow like tiny diamonds against the moonlight.

"Dad," she said, "how come no two snowflakes are alike, but they all have six points to them?" The question was not new, of course. Kepler had pondered the shape of snowflakes in the early 1600s. But this was my then six-year-old, stumbling upon a fundamental fact: symmetries may be manifest in many of the things we see, but they alone don't produce Nature's striking diversity.

"Snowflakes are kind of like people," I answered. "Even if we all have two eyes, two legs, and one head, we are all different. And it's the differences that make life interesting. Can you imagine if we all looked exactly like? If you looked just like me?"

"Yuk, Dad!"

"I thought you wouldn't be too thrilled."

That winter, it became clear to me that scientists and seekers of perfection from all walks of life have been courting the wrong muse. It is not symmetry and perfection that should be our guiding principle, as it has been for millennia. We don't have to look for the mind of God in Nature and try to express it through our equations. The science we create is just that, our creation. Wonderful as it is, it is always limited, it is always constrained by what we know of the world. And since we cannot know all there is to know, our science will always be incomplete. We may search for unified descriptions of natural phenomena, and we may find some partial unifications along the way. But we must remember that a final unification is forever beyond our reach. Just as a fish cannot conceive of the ocean as a whole, we cannot conceive of the totality of Nature. The notion that there is a well-defined hypermathematical structure that determines all there is in the cosmos is a Platonic delusion with no relationship to physical reality. It's an attempt to find God, even if metaphorically, through the lenses of science. There is only so much information we can gather. The human understanding of the world is forever a work in progress. That we have learned so much, speaks well of our creativity. That we want to know more, speaks well of our drive. That we think we can know all, speaks only of our folly.

PART IV

The Asymmetry of Life

36 LIFE!

There were no witnesses to what was about to happen. There was water and there was earth, and inert matter rained from the skies. Heat scorched the land and dried the seas, making the world glow. Smoke choked the air. What air was this? What was its composition? What kind of matter filled the land? What kind of liquid filled the seas?

Earthquakes rumbled across the soil, mixing dirt with volcanic gases. Tidal waves ravaged the seas and the land. Amid the chaos, sometimes things quieted down. The Earth would then cool for a while and the oceans were at peace. Following their chemical affinities, water and the many materials combined. Molecules formed and broke down, reacting in myriad combinations. Out of the primal ooze, some acquired shapes and started to chain up, growing and combining fiercely with each other.

Suddenly, one molecule outgrew the others, absorbed others, bending and twisting into the shape of a ladder. After it grew enough, it unzipped itself, becoming two. These found their pairs and twisted again into two molecules, as the first one had done. Then two molecules became four, four became eight, and so on. Life had begun, or something like it.

Not like us. But us nonetheless.

This is the second creation story of our generation: life emerging on primal Earth, self-organizing from an energy-driven assembly of inert materials. When did it happen? How did it happen? Where did it happen? If it happened here, could it have happened elsewhere in the cosmos? Is there alien life? Is it intelligent?

As with the origin of the Universe, a question that only a few decades ago was considered beyond the reach of science, the origin of life and its possible existence elsewhere in the cosmos is now the subject of cutting-edge research.

There is a chain of events linking the emergence of life on Earth to cosmic history. For life to have developed here, there had to be a planet at an appropriate distance from its parent star with the right chemicals readily available. For the star to have been born, there had to be a cloud of hydrogen, an element made around the same time the photons of the cosmic microwave background started to roam across space some four hundred thousand years after the Bang. The Universe had to be expanding at an appropriate rate for matter—both of the dark and normal kind—to coalesce into clouds: too fast, and matter would dissipate into the cosmic emptiness; too slow, and the Universe would fold back upon itself. Once formed, dark-matter clouds had to gravitationally attract hydrogen-rich matter, triggering its coalescence into the first stars and galactic structures. Before that, matter had to exceed antimatter. Even before that the very young cosmos had to have inflated to stretch the tiny quantum fluctuations of a primordial scalar field into large concentrations that attracted swarms of dark-matter particles, funneling them into clouds. Inflation had to have started somehow, from the bursting forth of a cosmic bubble. This is a story of instabilities and imperfections molding matter into the elusive shapes of primordial life: the asymmetry of time and the asymmetry of matter are the preconditions for the origin of life.

Once life took hold on our planet some 4 billion years ago, its history became one with Earth's. Natural selection set the stage for its evolution. No plan; no planner. Just time, chemistry, geology, and the struggle for self-preservation. From unicellular to multicellular, from anaerobic to aerobic, there followed tremendous diversification. Genes combined, interlocked, mutated. Life evolved. And still evolves. Life is one with Earth. This is a central point.

Here is a brief history of Earth's origin as we understand it now: The Universe was about 9 billion years old. Our galaxy, the Milky Way, or parts of it, already existed and was about 8 billion years old. Stars were being born and others were dying. A giant cloud, consisting mostly of hydrogen and some helium, hovered in space, revolving alone in the empty vastness. Suddenly a nearby star went supernova, fusing heavier chemical elements out of hydrogen in a paroxysm of nuclear alchemy. It pulsed and heaved and collapsed and expanded until it finally exploded, spewing its entrails across interstellar distances. The shockwave hit the lone hydrogen-rich cloud, sprinkling it with the chemicals of life—carbon, nitrogen, oxygen, sodium, iron . . . and made it unstable: ever-attracting gravity forced the cloud to contract while spinning around itself. Elongation along the equator and flattening of the poles followed, as more material concentrated in the center. There, density and pressure mounted. Matter, mostly hydrogen, got squeezed. As temperatures climbed above a staggering 15 million degrees Celsius, nuclear fusion began: hydrogen became helium, liberating an enormous amount of energy in radiation and neutrinos. The newborn star ignited, our Sun. Around it, in a pizza-like disk, material coalesced and collected into planetesimals, the ancestors of planets. Those farther out, where it was cold, gathered nuggets of frozen gases and grew large: Neptune, Uranus, Saturn, Jupiter. Those closer in only had rocky stuff to build upon and grew smaller: Mars, Earth, Venus, Mercury. Leftover debris accumulated in strips about the young Sun, like ornamental belts: between Mars and Jupiter, the asteroid belt, the boundary between the rocky and gaseous planets; outside Neptune, the icy balls of the Kuiper belt, where Pluto, once called a planet but now only a dwarf planet, orbits, along with some short period comets; further out, the Oort cloud, the home of trillions of icy balls, the nursery of comets. Earth, "the third rock from the Sun," was auspiciously positioned. If it had been much farther it would have been too cold: if it had been much nearer it would have been too hot. Conditions were just right for water to exist in its liquid phase, the cradle of life's chemistry.

The origin of life is a complex, multifaceted question. We will need to break it down into parts, each with its own set of remarkable ideas and possibilities. As we shall see, imperfections will play a central role in this story. The great challenge to understanding is one familiar to cosmologists: there were no witnesses to what happened. We must hunt for fossils, for clues that may open paths to the distant past, a time beyond our direct probing. We need to reconstruct a history knowing only too well that it is impossible to reconstruct it in every detail; we will never know *exactly* what happened 4 billion years ago in our young planet. As we argued before, we only know what we can measure and what we can measure is limited. But thanks to human inventiveness we can, in fact, know a great deal about our origins.

Scientific inquiry allows us to construct viable hypotheses about what *had* to happen for life to have emerged here. Detailed laboratory analysis of ancient geological samples, together with studies of possible biochemical pathways and computer modeling, work in tandem to pry open forgotten times. Astronomical observations reveal worlds old and new, teeming with possibilities. We may not be able to go back to Earth's infancy, but we can see new stellar systems and protoplanetary disks being formed in our cosmic neighborhood. We can detect extrasolar planets rotating about other stars and learn from their orbits. Even the chemical compositions of their atmospheres are becoming amenable to study. This way, we can search for telltale signatures of life, at least life as we know it, in worlds tens of light-years away from us. The astronomical search for other Earths is a search for our own origins and destiny. We are looking out to space to discover who we are. As we will see, what we have found thus far is enough to transform the way we think about ourselves and our planet.

37 THE SPARK OF LIFE

Luigi Galvani liked electrocuting frogs. Not out of sick sadistic pleasure, but out of scientific curiosity. The frogs were already dead and dissected, their muscles and nerves exposed. During the 1780s, in his laboratory at the University of Bologna, Galvani and his assistants performed a large number of experiments investigating the role of electricity in triggering muscular contraction. This was a time when the fascination with electricity was spreading across Europe and the Americas. In 1767, Joseph Priestley published *History and Present Status of Electricity*, describing what was known on the subject at the time. In the book, Priestley relates a series of gruesome experiments on the effects of electric shocks on mice, cats, frogs, and other animals. He was mostly concerned with the amount of electricity they tolerated and on their recovery (or not) from the discharges. It is possible that Priestley's investigations inspired Galvani. Priestley's book also included the first detailed description of Benjamin Franklin's famous kite experiment with the suggestion that lightning was but a huge electric discharge.

Investigations involving the rubbing of furs, hair, amber, and other materials indicated that apart from stormy clouds, electric charge seemed to pour out of matter's microscopic confines. All that was needed was an imbalance, a mechanism to generate an excess of charge in one region over another, and electricity would flow until equilibrium was reached. This electricity could be harnessed in devices known as Leyden jars, the precursors of the ubiquitous capacitor of modern electric circuits. Two separate metal foils coated the inside and outside of the glass jars. Using spark-making (aka electrostatic) machines, electric charge was transferred to the metal foils, where it remained "stored." Connecting the inside and outside metal foils would produce a sudden electric discharge that could be used in a variety of experiments. Electricity's mysterious origins and its potentially deadly effects fascinated natural philosophers and people alike. Parties where children and dogs received huge shocks from Leyden jars were quite fashionable in the late 1700s.[1]

In one of his experiments, Galvani observed that if a scalpel touched a frog's exposed sciatic nerve when an electrostatic machine sparked at the opposite end of the table, the frog's leg would twitch. If the sparks were made to fly off rhythmically, the leg would twitch rhythmically as in a dance. Excited, Galvani hooked the spine of a dead frog with a copper wire and hung it from an iron railing. Again the frog twitched as if animated with life. In an experiment performed during an electric storm, an ensemble of dead frogs hanging from a wire danced together whenever lightning struck. What a macabre spectacle! Electricity could make dead frogs into a chorus line! Galvani concluded that "natural" (from lightning) and "artificial" (from machines) sparks and the adjoining of different metals activated some sort of innate "animal electricity" residing in animal tissue. He conjectured that nerves were the conduits of this form of electricity, ultimately responsible for muscular motion. Electricity was the essence of life. To be alive was to be in a state of electric imbalance. Only in death would the electric activity cease.

A frenzy of experiments ensued. Cures based on electric shocks rapidly became the vogue for people suffering from all sorts of paralytic woes.[2] The word *galvanism* was coined to describe such electric manifestations in animal tissue. The wordsmith was none other than Alessandro Volta, Galvani's contemporary and academic adversary. Although they seemed to have treated each other with exemplary collegiality, they disagreed on the nature of the electric impulses flowing through the test subjects. Contra Galvani, Volta correctly asserted that nerves were the conduits of normal electricity, the only one that exists, as Franklin had previously established. There was no such thing as "animal electricity," only the flow of electric "fluid" through nerves (electrons were not discovered until 1897).

Volta used Galvani's discovery that two adjoining metals create a flow of electric "fluid" to invent the device we now call a battery. He piled alternating copper and zinc disks separated by cloth soaked in salt water and showed that an electric current would flow in a wire connecting the top and bottom of the pile. The wondrous invention, soon to be called a voltaic cell, supplied a continuous current that greatly facilitated research in electricity, opening a new field of inquiry to natural philosophers interested in the mysterious electrical properties of matter. What secret powers lay hidden in matter's microscopic structure? What relation did electricity have to life and death? Was life a state of electric imbalance? If nerves were the conduits of electricity and they bifurcate from the brain like branches in an upside-down tree, how did the brain create electricity in the first place?

These were some of the cutting-edge research questions of the early 1800s. There was much promise and fear. Should scientists probe into the elusive boundary between life and death? Could mortality be conquered? Or are certain questions beyond the province of man's inquiry? Should limits be imposed on scientific research?

On March 6, 1815, seventeen-year-old Mary Godwin lost her premature baby daughter, born a few weeks early. Visions of the dead baby haunted her for months thereafter. In a dream, Mary saw her dead daughter brought back to life after being rubbed vigorously in front of a fire. She was then living illicitly with married poet Percy Shelley and had had to endure his joy at becoming the father to a baby boy he had conceived with his now-estranged wife. Talk about adding insult to injury! Still, her love for Shelley only grew stronger and the couple had a baby boy early in 1816. That summer, they vacationed near Lord Byron's villa at the shore of Lake Geneva, in Switzerland. Dreary rain confined the friends indoors for days in a row. They talked of galvanism and of Erasmus Darwin's (Charles's grandfather) claim (or was it a reputation?) of having reanimated dead matter. As a pastime, Byron suggested that they each write a ghost story. Mary recalled how she took upon herself to fulfill the task: "I busied myself *to think of a story*. . . . One which would speak to the mysterious fears of our nature, and awaken thrilling horror." However, try as she might, Mary couldn't come up with a storyline. One night, as she lay awake in bed, she had a vision:

> I saw the pale student of unhallowed arts kneeling beside the thing he had put together. I saw the hideous phantasm of a man stretched out, and then, on the working of some powerful engine,

> show signs of life, and stir with an uneasy, half vital motion. Frightful must it be; for supremely frightful would be the effect of any human endeavor to mock the stupendous mechanism of the Creator of the world. His success would terrify the artist; he would rush away from his odious handiwork, horror-stricken. He would hope that, left to itself, the slight spark of life which he communicated would fade.[3]

Hence was born one of the great classics of literature, *Frankenstein: Or, the Modern Prometheus,* published in 1818. The novel is a far cry from the version made famous by Hollywood's 1931 movie, with Boris Karloff playing the "creature." Mary Shelley's "hideous phantasm" was no mentally handicapped, monstrous murderer. Albeit made out of scattered body parts animated by the "spark of life," the monster was a highly refined creature, who taught himself how to read and write. All he wanted from his creator was a mate:

> I demand a creature of another sex, but as hideous as myself. . . . It is true we shall be monsters, cut off from all the world; but on that account we shall be more attached to one another. Our lives will not be happy, but they will be harmless, and free from the misery I now feel. Oh! My creator, make me happy . . . do not deny me my request!

Terrified that he would create a race of monsters, Victor Frankenstein refuses. The story then turns tragic. Its subtitle tells it all. Immortal Prometheus stole fire from Zeus and gave it to the mortals. For this crime, he was chained to a rock and had an eagle eat and re-eat his liver for all eternity. Mary Shelley wrote a cautionary science-fiction tale, which rings especially true today with the ongoing debate on the ethical implications of genetic engineering, the modern version of cutting-edge science with the power to control life.

I was fourteen when I read *Frankenstein* for the first time. I can still feel my anger at the doctor for having forsaken his creature. The monster did not ask to be created, to look hideous. A scientist cannot neglect his creation, ever. And a scientist should always be prepared to face the moral and ethical implications of his work, be they good or bad. The uses and abuses of nuclear power—a truly dangerous kind of spark—are the most obvious example. Much has been written about this issue, and here is not the place to dwell on it. Science doesn't create or destroy. We do.

Reading *Frankenstein* as a teenager incited even more my fantasy of becoming a Victorian natural philosopher lost in the late twentieth century. When I joined the physics department at the Catholic University of Rio in 1979, I was the perfect incarnation of the Romantic scientist, beard, pipe, and all. I remember, to my embarrassment, my experiment to "investigate the existence of the soul." If there was a soul, I reasoned, it had to have some sort of electromagnetic nature so as to be able to animate the brain. Well, what if I convinced a medical facility to let me surround a dying patient with instruments capable of measuring electromagnetic activity—voltmeters, magnetometers, etc.? Would I be able to detect the cessation of life's imbalance, the arrival of death's final equilibrium? Of course, the instruments had to be extremely sensitive so as to capture any minute change right at the moment of death. Also, for good measure, the dying patient should be on a very accurate scale, in case the soul had some weight. I remember explaining my idea to a professor, possibly the same Professor Carneiro who had suggested that I read Weinberg's *The First Three Minutes* for my electromagnetism class. I can't remember exactly what he said, but I do remember his expression of muted incredulity.

Of course, I was only half serious in my excursion into "experimental theology." But my crackpot Victorian half, I am happy to say, had at least one predecessor. In 1907, Dr. Duncan MacDougall of Haverhill, Massachusetts, conducted a series of experiments to weigh the soul. Although his methodology was highly suspicious, his results were quoted in the *New York Times:* SOUL HAS WEIGHT, PHYSICIAN THINKS, read the headline. The weight came out at three-quarters of an ounce (21.3 grams), albeit there were variations among the good doctor's handful of dying patients. For his control group, MacDougall weighed fifteen dying dogs and showed that there was no weight loss at the moment of death. The result did not surprise him. After all, only humans had souls.[4]

38 LIFE FROM NO LIFE: FIRST STEPS

Frankenstein capitalized on the very effective combination of our fear of dying and the widespread belief that science's ability to pry open some of the deepest mysteries of Nature would eventually lead to the end of religion. Although I am not aware of serious experiments attempting to bring the dead back to life with electricity, I suspect there have been quite a few, even if undocumented. In fact, every time a stopped heart is made to pump again with the help of an electric discharge from a defibrillator we hear echoes of galvanism.

It turns out that the place to look for the deep relationship between life and electricity is not at life's end—in the reanimation of the dead—but at its beginnings. In 1952, a chemistry graduate student at the University of Chicago asked his Nobel laureate advisor for permission to set up an ambitious experiment: Stanley Miller wanted to reproduce the atmosphere of *prebiotic* (before life) Earth in a test tube and zap it with electricity. The sparks were meant to simulate the intense lightning during storms and active volcanism, believed to have been frequent in early times. In the flask (it was much bigger than a test tube), Miller put the chemicals that Harold Urey, his advisor, believed the atmosphere of early Earth should have: water, ammonia, methane, and hydrogen, then sent a discharge through them.* After a few days, Miller noticed an orange gooey mix accumulating in the bottom of the flask. To his excitement, he

* Recall the chemical compositions of these molecules: water (H_2O) is made of hydrogen and oxygen; ammonia (NH_3) of nitrogen and hydrogen; methane (CH_4) of carbon and hydrogen.

verified that about 10–15 percent of the original carbon from the methane was now in *organic compounds*, that is, chemicals which contain carbon and that form the majority of living creatures. These included nothing less than nine amino acids, the building blocks of proteins. By zapping a prebiotic soup of inorganic compounds with electric discharges, Urey and Miller created some of the stuff of life. Not life itself, to be sure, but the compounds that make life possible. Definitely a step in the right direction.

The experiment inspired many others, all with huge implications for our understanding of the origin of life. As more sophisticated research led to changes in the conditions believed to be prevalent in early Earth, so did the compounds used in Miller-Urey-type experiments: some took away the rich, energy-giving methane and the ammonia, substituting them with carbon dioxide and pure nitrogen gas and found no amino acids. Others added sulfuric compounds, released abundantly during volcanic eruptions, and found plenty. When ultraviolet (UV) radiation was substituted for electric sparks, amino acids and other organic compounds were still created, albeit with differing yields.[5]

Taken together, these experiments led to an extremely important conclusion: it is possible, starting with non-life-related (abiotic) inorganic chemicals, to create life-related (biotic) organic compounds, the first step in the complex transition from inanimate to living matter. This process is called *abiogenesis*, life from no life. Discounting supernatural tinkering, there is no other way. We now understand that the chain of events leading to life starts in space: chemical elements fused in dying stars make inorganic molecules that can then be synthesized into organic molecules. These organic molecules, if in the right planetary environment, grow in complexity, until some work together to become a living entity, capable of metabolism and reproduction. Imbalance is the driving force from link to link along life's chain: matter interacts and grows in complexity to neutralize charge asymmetries.

How and where did this life-from-no-life process happen? Has it been repeated elsewhere in the Universe? Or is Earth special? Organic molecules exist not only here but also in outer space: dozens of amino acids have been found in meteorites that have fallen on Earth; about 140 organic molecules have been spotted floating in interstellar space, presumably fused by the UV light radiating from young stars, like the UV variations of the Miller-Urey experiment. Outer space is a crucible for life's precursors. We should thus consider the very real possibility that the fundamental building blocks of life, especially amino acids, rained down on us from the skies. Of course, if life assembled itself here, using the drops of this cosmic organic rain, it could have done the same elsewhere. More radical proposals suggest that ready-made *living* entities could have rained down on us. This is the notion known as *panspermia*, that primitive life or/and its precursors travel across the cosmos as seeds blown by winds. If the seeds (sometimes called spores) happen to fall in a world offering the proper environment, they could germinate. Is the cosmos teeming with life or are we alone? The answer to this question may surprise you. And it will force you to rethink who you are and what the future holds for you and the rest of humanity. Taking all this in stride, let's start from the beginning.

39

FIRST LIFE: THE "WHEN" QUESTION

Dating when life first appeared on Earth is far from trivial. Organic matter has the inconvenient property of rapidly degrading after death. There is a reason why we quickly bury or burn our dead. Fossils of prehistoric animals, which are petrified organic remains or impressions cast in rock, need to survive the continuous grinding and shaping of time to be useful. Since geologists tell us that the history of ancient Earth is recorded in rocks, the most reliable information about early life is found through the detailed analysis of rock strata. This approach works well when we look for fairly recent animal remains, say, less than half a billion years old.* The problem is that early Earth, in particular during its first half-billion years or so, was a veritable hell. Planet formation is not the smoothest of processes: the young solar system was jam-packed with debris. While some of it accreted into the growing *planetesimals*, some settled into more or less stable orbits about the Sun. There were rocky and icy balls of sizes varying from a few miles across to mini-planets. Bombardments were the rule, as the craters on the surfaces of Mercury and of our own Moon show. In fact, modern theories of the origin of the Moon trace it to a giant collision between the Earth and a Mars-sized planet around 4.5 billion years ago, and so right after the formation of the Earth. It is supposed that the lopsided impact dislodged a huge amount of material that collected in a disk around the Earth. In time, the material in the disk coalesced into our silvery satellite. In spite of the apocalyptic violence of the event, there is something quite lyrical in the thought that the Moon is truly Earth's

* Recall that the Earth is approximately 4.6 billion years old.

daughter, duly mixed with the remains of whatever celestial body collided with us. The Moon is Earth's yanked rib.[6]

The net result of these violent impacts was a young Earth frequently tossed into a molten state. Rocks and metals were not able to remain solid for very long; as soon as a quiet period allowed for some crustal cooling and solidification, another major impact would set the whole planet or at least large parts of it back into a bubbling soup of lava. Water, most probably present in primitive oceans, would have been for the most part vaporized into the atmosphere. Under these circumstances, life, or at least sustained life, was impossible. Even if molecular assemblies started to stir in the direction of organic complexity during periods of relative quiet, their bonding efforts wouldn't have lasted long. To make this period of Earth's history more opaque, rocks would not have been there to record anything. The exact course of events, including the possibility that life somehow managed to emerge under such conditions, will probably remain unknown.

The planet's youthful excitement had to subside eventually. Although details about the precise timing of the cosmic bombing slowdown are still under debate, it is fairly safe to say that soon after 3.9 billion years ago things got quieter.* Earth cooled; rocks solidified somewhat; water, minerals, and simple organic compounds collected into shallow ponds; and the prebiotic soup started to brew in earnest. Molecules drifted and reacted with one another, in a dance of electric attraction and repulsion: life's choreography had begun. Clay surfaces, tidal ebbing and flowing, and underwater hot vents could also have enabled life-building reactions. (The "where" question will be addressed soon.) How long did life take to emerge from a prebiotic soup? As the late biochemist Leslie Orgel once told me,

* Recent studies based on the analysis of lunar rocks and models of the solar system formation suggest that a "late heavy bombardment" took place around 3.9 billion years ago. If confirmed, this late bombardment would make it even more remarkable that life could have formed so soon after 3.9 billion years ago. Of course, if things were relatively quiet *before* the bombardment *and* water was abundant, it is possible, although unlikely, that life formed quite early, and that some of it survived the chaos. In this case, life would be older than we think. Another possibility is that life formed and was extinct several times before finally taking hold sometime after 3.8 billion years ago. Whenever life originated, we still need to understand how.

"That's not the right question; life emerged immediately. The question is, how soon in Earth's history did life take hold?" So the "when" question can be phrased thus: How soon after the bombardment subsided can we find the first signs of persistent life?

Starting with incontrovertible evidence, signs of life are found at 2.8 billion years ago. These are colonies of stromatolites, layered mushroom-shaped brown rocky structures formed in shallow water from the fossilized remains of microorganisms. Evidence indicates that the microorganisms forming the colonies were cyanobacteria (aka blue-green algae), capable of producing oxygen via photosynthesis. The advent of cyanobacteria in early Earth is believed to have radically changed our planet: they are largely responsible for making energy-giving oxygen into an abundant atmospheric gas, with tremendous implications for the complexity of life. But more on that later. Right now we are interested in going backward in time. Stromatolites found in Australia and dated at 3.5 billion years old are widely accepted as having signatures of early life. Original claims that these were also due to cyanobacteria are now less favored; an alternative explanation is that these are the remains of some primitive microbe that lived in the surroundings of a hydrothermal vent. In either case, present evidence allows us to push the origin of life to 3.5 billion years ago. Can we go earlier?

Recent claims of primitive life in rock samples from Akilia Island in western Greenland, dated back to 3.85 billion years ago, are still under much dispute. Here the evidence is based on the ratio of two isotopes of carbon, C-12 and C-13. (So, one with six and the other with seven neutrons in the nucleus.) An excess of C-12 relative to the heavier C-13 often indicates the presence of biological activity: biochemical processes like to save energy and work with the lighter of the two. And just such an excess seems to have been measured in the available samples. Since other possible explanations independent of organic life exist, the question is not yet settled. If proven correct, the rocks from Greenland would push the origin of life tantalizingly close to the end of the bombardment era, strengthening the argument that life emerges very fast once it's given a chance. This being the case, life should be quite widespread in the cosmos.

In 2008, claims of an even earlier life marker were made public. The samples, now from Western Australia, are from such a primitive time in Earth's history that no rocks could have survived. Instead, the isotopic excess of carbon mentioned above was found in minute pieces of graphite and diamond locked inside extremely hard zircon crystals dating back to 4.25 billion years ago. Again, other possible abiological mechanisms may explain the isotopic excess. At the time of writing, it's prudent not to consider either claim as being conclusive but to set the clock for first life instead at *no later* than 3.5 billion years ago. Life could have originated before that, but at this point we don't know for sure. It could even have been originated several times and destroyed several times, leaving no detectable trace. We will probably never know what exactly happened in these early times. Such are the limitations of science: we only know what we can measure. However, with or without a late heavy bombardment, the reliable evidence at hand proves that primitive life emerged within a few hundred million years of relative quiescence, not much compared to Earth's age of 4.54 billion years. First life was nimble.

40

FIRST LIFE: THE "WHERE" QUESTION

When I was fifteen years old, my older brother Luiz took me to his chosen nature sanctuary, the then almost deserted tropical island of Itacuruçá, about ninety minutes south of Rio. The small island is one among hundreds that line the coast between Rio and São Paulo, right around the Tropic of Capricorn. The young Charles Darwin captured the magic of the region perfectly in the entry dated April 8, 1832, in his *Voyage of the Beagle:*

> The view seen when crossing the hills behind Praia Grande was most beautiful; the colors were intense and the prevailing tint a dark blue; the sky and the calm waters of the bay vied with each other in splendor. After passing through some cultivated country, we entered a forest, which in the grandeur of all its parts could not be exceeded.

"Most beautiful," "splendor," and "grandeur" are the right words to express the spectacular confluence of clear tropical waters and luxuriant jungle that graces these parts. Life explodes from everywhere: trees draped in orchids and vines, their branches bent heavy with flowers and exotic fruits; birds of all sizes and colors, a profusion of spiders and beetles, of ants with huge red heads and with tiny ones, and fish, shrimp, and crabs of all kinds in the waters. At night, the eerie silvery light of a luminescent plankton the locals call *argentia* animates the gentle rocking of the waves. The sea heaves like a giant living being.

"Here, Marcelo. Sit on this rock and contemplate for a while. Close your eyes; open your soul. Feel life!" said Luiz.

I remember the view from that rock—"Pedra da Baleia," Whale Rock—as if it were yesterday. I wish it were. A large, round granite boulder, the rock stood some fifteen feet above the sand, facing the many islands just hundreds of yards away. Above it, acacia branches made a makeshift roof of bright yellow flowers. A school of robalos hunted for shrimp in the distance, roughing the surface of the water. A Bem-te-vi ("I saw you!"), the signature bird from tropical Brazil, called its mates on and on with its mantra-like tune. I was enchanted. After a few moments of contemplation, I felt my inner self dissolving into the scenery. Life's powerful force took hold of my whole being. I experienced what to me was an entirely new dimension of the sacred, a deep sense of communion with life at a scale way beyond the human, what mystics often refer to as the *numinous*. Even today, decades after those brief transcendent moments, a shiver runs down my spine when I close my eyes and try to re-create that magical scene. Life worships at the shrine of Nature. Science, at least the way I see it, is one of the doors that lets you into the temple.

My brother brought me back to Earth. "Marcelo, enough of this. Let's go for a walk." We took a narrow trail that circles the island, cutting through beaches and forest, offering a view not too different from that which Darwin witnessed long ago. At some point, we stopped to investigate a muddy puddle, filled with a gooey, Jell-O-like, yellowy substance. "This" said Luiz, "is the stuff of life. It all came from something like it." I poked at it with a stick, trying to imagine molecules coming together to become a living thing. "How do you know when something is alive?" I asked. "Good question," my brother replied. "Perhaps one day you can try to figure it out." It took me many years, but here I am, brother, trying to figure it out.

Embedded in any discussion of life's origin is a definition of life. Not an easy thing to do. There is no universally agreed upon definition. There are what scientists call operational definitions, good enough to get us going. For example, "Life is that which is squishy." Not very scientific, but it works in many occasions, especially in playgrounds and summer camps, when children go around collecting bugs or running away from bees. Going a bit deeper, we can say, "Life is a chemical reaction network that extracts sustenance from an environment and reproduces." This one is better, but not all living things reproduce, as we know from babies and elderly people, or from fertility clinics. So, let's refine the definition: *Life is an autonomous chemical reaction network capable of extracting sustenance from an environment and equipped with the ability to reproduce*. This definition, although not perfect, works for our purposes. Viruses and bizarre prion proteins are not included, being relegated as mere *replicators*, since they can only reproduce after taking over the replication materials of a host cell or of proteins, respectively. They are only alive inside living things. In searching for life's origins, I take the logical point of view that the focus should be initially on the hosts and not on the parasites that need living hosts to exist in the first place.

The definition in italics implies that life is a highly constrained system, even if we push its possibilities to the limits of plausible biochemical processes. First of all, life needs the right chemicals. To understand the origin of life, we have to figure out what these chemicals are and where they came from. This is related to the "how" question, to be addressed next. Second, life needs the right environmental conditions. Specifically, life as we know it needs liquid water and warmth. Chemical reactions occur when the various atoms and molecules diffuse and meet each other. Put another way, we can think of atoms and molecules as small bits of matter where positive and negative electric charges are not always distributed symmetrically. Given the chance, these asymmetric electric bundles will connect with each other in an attempt to minimize these charge asymmetries. Water is the medium—the universal solvent—allowing for these atomic and molecular encounters. Since water can only be a liquid within a certain temperature and pressure range, life is possible only within that same range. At ground level here on Earth, the range is roughly between zero and 100 degrees centigrade (between 32 and 212 degrees Fahrenheit), when water is in its liquid state. Some bacteria may actually survive and grow below 0 degrees C, while "extreme thermophilic" ones can survive in temperatures of up to 85 degrees C. In rocks deep below the surface or under the ocean, where pressures are higher, bacteria may grow at temperatures above 100 degrees C. In spite of this bacterial bravado, a temperature of 115 degrees C seems to be an absolute upper limit for life. High temperatures tend to break the molecular bonds that make life possible: atoms are freed and go their own ways.

Of course, it is conceivable that life can exist without water as a solvent or without a carbon-based chemistry: life as we don't know it. Maybe the solvent could be liquid ammonia or the basic element could be silicon. Even so, the temperature range for life would not be very different from what we stated above.*

These arguments make the "where" question more approachable. Even taking into account extremophiles, microorganisms that live under extreme conditions of temperature, acidity, alkalinity, and pressure, the first life would have needed liquid water. However, too much liquid water dilutes the concentrations of the chemicals, making it hard for them to find each other and interact. Maybe shallow pools provided a high enough concentration of chemicals and all went according to plan: reactions started and molecules grew in complexity until they became a self-sustaining reaction network, extracting energy from the environment to keep on going. At some point, a membrane grew around a group of reacting chemicals, creating a *protocell,* a primitive cell. Isolating these chemicals behind a semipermeable membrane allowed reactions to thrive. Thus was born the first prokaryote, the first single-celled organism.[7]

In a 2009 paper with my then graduate student Sara Walker, we showed that it is possible, at least in principle and within very simple chemical networks, to go from pure chemistry to a reactive system separated from the outside by a simple membrane. Although our theoretical model is far from creating a living cell from chemistry, it presents plausible steps going from chemistry to biology, what can be called a "bottom-up" approach to the origin of life. Since at its most fundamental level biology is, in a sense, living chemistry, self-organizing chemical networks must have hit upon the cell membrane as the best way to kill two birds with one stone: protect themselves from attackers while being able to extract energy and nutrients from the environment. A castle with thick walls but no windows or doors is a strong fortress but the people inside won't survive very long. A castle with flimsy walls will easily be overtaken. The best castles have an architecture that allows for a balance between protection and access. Cell membranes operate the same way.

* Since ammonia boils at −33.34 degrees C, in order to keep it liquid the temperature has to be sufficiently low or/and the pressure sufficiently high. Silicon, like carbon, forms four bonds with other atoms. However, the larger silicon atom is more vulnerable to chemical attack. As a consequence, silicon-based biochemistry is quite poor compared to that based on carbon.

During the past decades, scientists proposed many possible venues where primitive reactions leading to first life could have taken place. Clays were probably abundant in early Earth and could have served as templates for the reactions, facilitating certain kinds of chemical bonding. We can imagine shallow lagoons rich in organic molecules evaporating and leaving behind a high concentration of organic materials in the clay. Also, tidal ebbing and flowing could regularly do both the wetting and the drying needed for the reactions to evolve toward higher complexity. But don't think of tides in early Earth as a gentle process. They were much more dramatic than they are today. Recall that tides are caused by the mutual gravitational attraction between the Earth, Moon, and Sun. Since the gravitational force weakens with the square of the distance and since, at early times, the Moon was much closer, we know tides were much stronger.* If we think of the Earth as a sphere covered with water, the side facing the Moon, being nearer to our satellite, experiences a stronger attraction than the center of the Earth or its opposing side. Even though water, being more malleable, is deformed the most, Earth's surface also gets pulled toward the Moon. Today, because the Moon is farther away and the Earth is harder, the crust deforms by only about twenty centimeters and tides reach an average of three-quarters of a meter (a little under two and a half feet). But in the first hundreds of millions of years of Earth's existence, when the Moon was much closer and the Earth was still molten, its crust raised and sank by over sixty meters (about two hundred feet) and the oceans (if present) by two hundred meters at each tide! As time passed, the Moon moved outward and the Earth cooled off and hardened.[8] Still, at about the time when we expect first life to have formed, between 3.8 and 3.5 billion years ago, tides lifted the oceans by many meters. Early islands would be periodically submerged.

Meanwhile, deep in the oceans, hydrothermal vents spilled hot material from Earth's interior into their surroundings. The discov-

*Incidentally, the huge early tidal deformations on both the Moon and the Earth dissipated much of their initial rotational energy. (Part of the energy used to deform the Earth and Moon comes from their rotational motion.) As a result, Earth's rotation slowed down to today's value of a revolution in just under twenty-four hours, while the Moon, being less massive, slowed down to a halt, and has shown the same face to us ever since.

ery that nowadays life is abundant around submerged, dark, and oxygen-deprived volcanic cones has inspired many to consider the possibility that it may have first originated around such extreme environments. For this to happen, the right chemicals must have been present in the right concentrations. The idea is tantalizing. Even if this was not the path that life took on early Earth—and there seem to be practical difficulties with this scenario, including achieving high enough concentrations—life may have arisen this way, or in some way like it, elsewhere in the cosmos.

The fact that life is so resilient, able to thrive in such hot places in the absence of light and oxygen, opens the possibility that it may be more universal than we could have guessed even a few decades ago. Although Darwin's "warm little ponds" are still the easiest places to conceive of the first life, current research has shown that we should be open to unexpected surprises. Quite possibly, life emerged in different places under different conditions. If we expect life to exist elsewhere in the Universe, where conditions were and are certainly different than they were here, this "many-venue" approach is almost a given. We should think of the many possible life-originating venues on Earth—warm coastal ponds, clay minerals, marine hydrothermal vents, sea-ice-forming regions—as small-scale laboratories for the types of life that may exist elsewhere. There is not *an* origin of life but *many* plausible origins of life.

We now take leave from the "where" question by quoting from Erasmus Darwin's *The Botanic Garden,* first published in 1791. Young Charles surely read his grandfather's famous poem published seventeen years before his birth:

> Organic Life beneath the shoreless waves
> Was born and nursed in Ocean's pearly caves;
> First forms minute, unseen by spheric glass,
> Move on the mud, or pierce the watery mass;
> There, as successive generations bloom,
> New powers acquire, and larger limbs assume;
> Whence countless groups of vegetation spring,
> And breathing realms of fin, and feet, and wing.

Evolutionary ideas clearly ran in the Darwin family.

41

FIRST LIFE: THE "HOW" QUESTION

Life is an excellent illustration of reductionism's limitations. Although every living being is ultimately an ensemble of atoms joined by chemical bonds, life defies this sort of description. To go even further and reduce life to elementary particles interacting via the four fundamental forces borders on the ridiculous. The Theory of Everything that Unifiers are wedded to has nothing to say about the workings of life. Granted, most of them would be the first to acknowledge this. The road from particles to atoms to molecules to huge biomolecules to metabolizing and reproducing chemical networks is highly disjointed, a point that Nobel Prize winner Philip Anderson, biologist Stuart Kauffman, and many others have made quite clear over the past years. The techniques used in quantum mechanics, successful as they are in describing the behavior of electrons and the simplest atoms and ions, fail as we move to larger atoms and complex molecules.*

As we have seen, atoms and molecules bond electrically: some have charge to give or share (donors), while others need charge (receptors). Bonding softens large charge imbalances by decreasing the energy of a system: as with income taxes, where married couples get the best rates, it is often energetically more advantageous to be together than to be apart. (Of course, chemistry doesn't discriminate among legally and not legally married atoms as the IRS does with couples.) In a general sense, chemistry describes matter's urge

*The fault is not entirely with quantum mechanics. As some of the more technically minded readers know, even in classical mechanics we can't solve for the motion of three or more interacting bodies exactly. This means that atoms with two or more electrons must be studied via approximation methods.

to bond as an attempt to decrease asymmetries in electric charge distributions. Life is a very complex manifestation of this urge, an imbalance that re-creates itself.

In Nature, things change to not change. A system in stable equilibrium, where attractions and repulsions are balanced, doesn't change. Even if it does undergo local fluctuations, on average it stays the same. You may twitch about your chair while reading this book, but unless you decide to stand up you will remain seated in the same spot, the local equilibrium point. More precisely, a system in stable equilibrium is immune to small disturbances: when perturbed, it always goes back to its stable position. (With a little help from friction. Otherwise, it would keep on oscillating about its stable point.) Think of a marble oscillating inside a soup bowl. After a while, it will stop at the bowl's lowest point. In contrast, unstable equilibrium leads to change; small disturbances may drive the system away from its initial position or state. If, instead, the marble is balancing on top of an upside-down soup bowl, it will take off with the slightest touch. You may also induce change by tossing a system in equilibrium into an out-of-equilibrium state. We do this every time we pour cold water into a hot bathtub, for example: the water will cool off to a new equilibrium state at a lower temperature. Whether starting from an unstable equilibrium situation (ball on top of upside-down soup bowl) or tossing a system into an out-of-equilibrium state (cold water into hot bathtub), disequilibrium leads to change. Some systems, like the stock market, are always out of equilibrium: stock values are always changing, either creating or destroying wealth. Living systems are also examples of permanent out-of-equilibrium systems. To keep on living, organisms need to absorb energy and nutrients from the environment, leaving behind their downgraded remains. For life, equilibrium means death.

One of the most revealing discoveries of modern science is that many complex patterns and structures that we see in Nature—galaxies, hurricanes, ocean currents, living beings—are mechanisms to counter imbalance. On the road from change to no-change, from disequilibrium to equilibrium, all sorts of wonderful (and terrible) things happen. For example, when a rock is thrown on a quiet pond, it transfers the energy of its motion to the water. Quite quickly, outgoing water waves dissipate this excess energy. The waves are coherent macroscopic structures that work to reduce an imbalance. As they do so, they eventually restore the pond to its equilibrium situation. In general, interactions between the many components of a system generate complex forms that work toward leveling things out. These include atmospheric temperature and pressure differences that generate winds and hurricanes, or a collection of stars that self-organize into galaxies, or excesses in the concentrations of chemicals in a solution that triggers life-sustaining reactions. Imbalance leads to change that leads to form that leads to balance. This is the essence of Nature's imperfect cycle of creation.

Under this view, a cell is a complex, partially isolated, self-sustaining chemical reactor whose main directive is to downgrade energy. In order to function, cells absorb high-grade usable energy from the environment and secrete it back in a low-grade useless form. (We do the same when we eat food and secrete our waste.) It follows that the more cells there are, that is, the more they reproduce, the more efficiently they will perform their energy-downgrading function. (Likewise, the more people exist, the more food is eaten and the more waste is secreted.) As a consequence, reproduction serves a very clear purpose: life makes more life so as to keep on downgrading energy. Life is a mechanism to decrease imbalances in the distribution of energy, a sort of steamroller that averages energy excesses away.*

Do not let this mechanistic perspective on the energetics of life discourage you. The wonder is to be found both in life's function, through its ingenious biochemical mechanisms, and in its form, through its remarkable diversity. As Darwin wrote in the closing of his *On the Origin of Species:* "from so simple a beginning endless forms most beautiful and most wonderful have been, and are being, evolved." This numinous dimension of life should inspire awe and celebration.

• • •

* And what energy excesses are these? Mainly, the energy output from the Sun that heats the Earth. When Earth absorbs a solar (visible) photon and later emits infrared photons to space (at an approximate rate of one to twenty) it is exporting a lower grade of energy. The energy difference in the downgrade (or, more technically, entropy increase) is used to feed organized structures on Earth, from hurricanes to life.

To understand life's simple beginnings we have to start with prebiotic chemistry and know which ingredients were available for bonding in early Earth. What chemicals bonded to emerge as an animated creature from an inanimate soup? How did nonlife become life? This fascinating, and very difficult, question has yet to be solved. We have seen how little we know of the prevalent environmental conditions on primeval Earth; different Miller-Urey experiments use different recipes for the primordial soup and for the all-important life-triggering spark. (Well, amino-acid-triggering spark, anyway.) Yet in the currently accepted choices, chemical mixing and bonding consistently lead to several important ingredients of all living beings.* As with models of cosmological inflation discussed previously, we may lack the specific details but we have a general framework.

Another view is that Earth may not have been the crucible for first life's ingredients: they may have rained down from the skies, either by direct deposition, as Earth's gravity captured rogue organic molecules flying across the interstellar void, or courtesy of meteoritic impacts. The discovery of many amino acids in some meteorites, most notably on the large one that fell on Murchison, Australia, in 1969, shows that many of the basic ingredients of life are synthesized in outer space. Indeed, the list of life-related (and unrelated) organic chemicals found in the Murchison meteorite is quite long.

Early in 2006, I interviewed Stanley Miller for a Brazilian TV documentary in his laboratory in La Jolla, California. He was recovering from a stroke and spoke with great difficulty. I confess my boyish excitement at being near the great man, touching the apparatus that made him famous. The brownish yellow, pre-life goo was very visible at the bottom of the flask. Miller pushed a button and sparks flew out of small electrodes. Images of Dr. Frankenstein's laboratory were hard to fight back. As I struggled to keep them to myself, I asked Miller what he thought of the panspermia idea. "Nonsense!" he exclaimed, agitated. "It all started right here."

One of the difficulties with the it-came-from-outer-space hypothesis is that organic molecules tend to be fragile. Recall how spacecraft shine aglow as they reenter the atmosphere. Similarly,

* This is true as long as the primeval atmosphere contains energy-rich, reducing (electron-giving) gases. Otherwise the products don't include amino acids.

molecules could break apart at entry into our atmosphere, although cosmic drifters with low mass and low entry velocities would rain down gently. If transported by a meteor, organic compounds could be destroyed either during entry or right after impact. Those who defend the alien organic seed hypothesis argue that samples found in meteorites such as the Murchison and others are proof that some chemicals survive entry and impact. Fair enough. Also, recent studies show that even if the outer layers of meteors get seared at entry, their cores can remain quite cool. Possibly, molecular hitchhikers nestled deep into the rocks arrived here safe and sound, although, of course, they still needed to find a way out of their pods. As with the "where" question, given the current lack of evidence, it's prudent to keep an open mind and to consider both mechanisms—made here or rained down from space—as viable. Perhaps they worked together to enrich the prebiotic soup. In any case, as with any recipe, the ingredients are only the first part. The "how" question has many stages.

42

FIRST LIFE: THE BUILDING BLOCKS

Let's go back to the time when first life took hold on Earth. Somewhere around 3.6 billion years ago (or earlier), possibly in a drying lagoon, a set of carbon-rich chemicals, including amino acids, reacted with growing complexity as they tried to minimize charge imbalances, creating longer and longer molecular chains. These chains combined with each other, self-organizing into increasingly more complex structures. Possibly, simple carbohydrates (foods) appeared as well. Then, somehow, these chains began to divide into imperfect copies of each other. We will never know with certainty which molecules made up these chains or how they developed into self-reproducing entities. All we can do is move backward and try to reverse-engineer plausible scenarios for first life from the life we can study in the laboratory today. Even if our first common ancestor, the first living being, has not left us any trace of its existence, we can still learn from what we know now and travel backward along the path leading to first life.

The simplest unit of life, the simplest living thing, is a cell. Now, cells themselves come in different types and sizes; they certainly evolved with time. A typical cell is about one-hundred-thousandth of a meter across (ten microns), about a tenth of the thickness of a (thin) human hair. Some cells are quite big, the largest being an unfertilized ostrich egg. Blue-green algae and many bacteria are prokaryotes, primitive cells where the genetic material, the DNA used in their reproduction, is bundled into a coil without a membrane separating it from the rest of the cell. In eukaryotes, the more sophisticated cells like the ones in our body, an isolated nucleus houses the genetic material. As we look into the history of life on

Earth, we discover that single-celled organisms were by far its most enduring inhabitants. The numbers are remarkable: from around 3.6 billion years ago, life remained unicellular until about 1.6 billion years ago. That is, for roughly 2 billion years or more, life on Earth consisted *only* of single-celled organisms, albeit some organized in colonies. Eukaryotes appeared close to the end of this period, when oxygen became more abundant in the atmosphere thanks to collective efforts of the photosynthetic blue-green algae.*

This fact should give us pause. To study life's origin we can forget about multicellular organisms. The stars are the prokaryotes. The crucial transition from single-celled to multicelled organisms, from our amoebalike ancestors to sponges, happened for a number of improbable factors: most importantly, the increase in atmospheric oxygen between 2.7 and 2.2 billion years ago. A consequence of this increase is the parallel production of ozone due to the action of UV sunlight on oxygen. This ozone created a protective layer between organisms and the same nasty UV radiation from the Sun, allowing more complex life forms to evolve. We wouldn't be here without it. When we later consider the possibility of life elsewhere in the cosmos these factors (and many more) will be crucial.

Back in our warm little pond, what compounds promoted the big jump into life? The honest answer is that no one knows. There are two competing views. One claims that metabolism came first, a view espoused by Alexander Oparin, a pioneer of origins of life studies, and, more recently, by the physicist Freeman Dyson and chemist Robert Shapiro. The other, more popular, view claims that genetics came first. Let's look briefly at both, preparing the ground for our discussion of the role of molecular asymmetries in the origin of life.

In his 1924 book *The Origin of Life*, Oparin noted that drops of oily liquids don't generally mix well with water, forming small bubblelike droplets instead. Anyone who has mixed olive oil and vinegar to make salad dressing has seen this. These fatty droplets, according

*These dates are approximate. There is still much debate as to when multicellular life—as distinguished from colonies of unicellular organisms—appeared, although it is certain that it was widespread by at least 550 million years ago, during the so-called "Cambrian explosion," a period of inordinate diversification of life forms. There are indications that sponges—the first multicellular organisms—may have been present as early as 1.8 billion years ago.

to Oparin, would have made a nice protective environment allowing molecules accidentally trapped in their interior to react with each other with reduced external interference. Occasionally, certain reactions would produce more chemicals and grow in complexity. At a critical threshold, the molecules would be able to produce more copies of themselves in a self-sustaining ("autocatalytic") reaction network: the little fatty bags would become the first protocells. As opposed to reproduction in a more organized genetic framework, reproduction here would initially happen at random, as turbulent external conditions would force some droplets to split. (Again, shaking salad dressing shows this.) In rare cases, the daughters would contain the right chemicals to also maintain self-sustaining reactions and a population of somewhat similar protocells would start to develop. Doron Lancet and collaborators at Israel's Weizmann Institute have developed complex computer simulations of such "lipid world" scenarios, showing that when a parent cell can produce more than one self-catalyzing daughter, a chain reaction may occur, leading to a kind of primitive life. Genetics would develop later, as the reproductive process perfects itself through countless "generations," led by the invisible hand of some prebiotic version of natural selection. We should expect that protocells containing molecules that reproduced more efficiently and could better extract and metabolize energy from the outside environment had an advantage over others and slowly came to dominate the population.

The countering position is that genetics came first: duplication preceded metabolism. The most popular idea within this view is the "RNA world" hypothesis: of the two genetic information carriers, DNA and RNA, RNA is the one with the ability to jump-start its own duplication process. Unlike DNA, it can function as an enzyme, so it's able to catalyze its own polymerization (that is, the chaining of smaller pieces into longer molecules like pearls in a necklace) and duplication. If we assume, quite reasonably, that life started simple, a self-sufficient replicator is one way to go.[9]

As Tom Fenchel remarked in *Origin and Early Evolution of Life*, the real advantage of the RNA-first scenario is that it allows for extensive laboratory-based studies. Many remarkable experiments, such as those by Manfred Eigen and Leslie Orgel and, more recently, by Gerald Joyce's group at the Scripps Research Institute in

San Diego, have clarified the relationship between genetics and natural selection at the molecular level through direct RNA and DNA manipulation, illuminating the connection between chemistry and biology. However, from the point of view of life's origins, it should be clear that for RNA to be present in early Earth, a lot of complex chemical syntheses must have taken place already. As Fenchel wrote, "It is obvious that the putative RNA world could not have started in a vacuum."[10] One difficulty, for example, is that Miller-Urey-type experiments have so far failed to produce nucleosides, the chemical bases such as adenosine and cytidine commonly found in RNA and DNA. If you can't make the bricks you can't make the skyscraper. But the situation may be changing. In May 2009, three British chemists from the University of Manchester reported a major advance in support of the RNA world. Using an innovative sequence of chemical reactions, the trio managed to synthesize two of the four nucleosides, bypassing many of the difficulties other groups encountered during the past twenty years. As an add-on, they used UV rays, which were abundant in prebiotic Earth, to speed up the synthesis. Also, reactions worked best at fairly high temperatures, around 60 degrees centigrade (140 degrees Fahrenheit). The discovery was quickly hailed as a huge step toward our understanding of how life originated on Earth. Even so, we must exert caution; the fact that scientists managed to achieve the synthesis of nucleosides in the lab following a specific pathway doesn't mean that Nature chose the same path.

There have been proposals where simpler organic molecules jump-started the replication process, such as peptides (compounds with two or more amino acids linked in a chain and joined with certain types of bonds). In spite of many interesting suggestions, we still can't be sure of the way Nature bridged the gap between inanimate and living organic chemistry. Quite possibly, as Dyson suggested in his *Origins of Life,* both scenarios worked to generate the first "thing" we could call living: at some point, protocells with primitive metabolism and simple lipid boundaries—the cellular hardware—were invaded by or accidentally absorbed the precursors of genetic replication—the cellular software—as parasites invade a host. After aeons of trial and error, a symbiotic fusion of the two eventually developed, creating a cell with optimized replication capability.

Although the search for the first replicator is a fascinating research topic, our interest here resides in the fundamental asymmetries and imperfections that are ultimately responsible for the complex forms found in Nature. We have discussed how the asymmetry of time is intimately related to the asymmetry of matter and how the structures that populate the cosmos, the luxuriant garden of galaxies and galaxy clusters, sprouted from seeds planted during primordial inflation. Having surveyed several questions and challenges related to understanding the origin of life, we are ready to extend our reach into the essence of life itself. As we will see, asymmetries at the molecular level play a fundamental role in the origin and evolution of life. From molecular structure to replication, without imperfection, life would be impossible.

43

THE MAN WHO KILLED THE LIFE FORCE

Whenever you drink pasteurized milk, thank Louis Pasteur for his tenacity and strict laboratory methodology. Thank him also for explaining how diseases originate from germs and for the production of some of the first vaccines, including that for rabies. As a by-product of his research, Pasteur realized that fermentation, such as that occurring in the making of wine and beer, is a biological process due to the presence of microorganisms. Leave it to a French chemist to refine the science of winemaking.

Pasteur delivered a decisive blow against the notion of the spontaneous generation of life, a view that had been prevalent since the times of Aristotle, and according to which live animals could emerge spontaneously out of dead matter. If the notion that mice may emerge from moldy grain, flies from rotten meat, and frogs and salamanders from mud sounds ridiculous to you, it didn't to most people up to the mid-1650s. Popular recipes explained how to produce bees by burying a young bull with its horns protruding out of the ground or mice by placing dirty rags in an open pot containing wheat. In 1668, the Italian physician Francesco Redi published the account of a decisive experiment, where he placed pieces of meat in several jars, leaving some to rot in the open while sealing others. Not surprisingly, maggots and flies were only seen in the jars left open. Redi correctly concluded that flies and maggots didn't generate spontaneously in the rotten meat but were transported through the air. However, with the invention of the microscope at about the same time, living creatures were seen to exist in the realm of the invisible. Supporters of spontaneous generation didn't waste any time. Could it be that bacteria did appear out of nothing and the process of spontaneous generation was just invisible to the eye? The argument raged on for decades.

Around 1750, the Scottish clergyman John Needham claimed to have demonstrated that air had an intrinsic life force capable of generating bacteria. In his experiments, he saw microorganisms appearing in soup left in open containers. He even saw microorganisms in soups that he briefly boiled and poured into supposedly clean flasks that he then closed with corks. Could there be a mysterious life force hidden in the realm of the invisible? Again, an Italian came to the rescue. During the mid-1760s, Lazzaro Spallanzani demonstrated that Needham's microorganisms would disappear if the soups were boiled long enough. Also, the corks were only partially blocking the air and hence allowed microorganisms to penetrate. Not to be defeated, Needham counterargued that Spallanzani's hour-long boiling "killed" the air's hidden life force. No truce was in sight.

The situation was so bad that, in 1860, nearly a century after the Needham-Spallanzani arguments and two centuries after Redi's experiments, the Paris Academy of Sciences offered a prize for an experiment that would resolve the dispute once and for all. Pasteur claimed the prize in 1864. His solution was as brilliant as it was simple: he designed flasks with very long S-shaped necks (swan-neck flasks) and boiled soup inside them, leaving them open. He also sealed flasks of varying shapes and neck length with cotton, as he wished to demonstrate that cotton could stop outside bacteria from coming inside the flasks. He quickly concluded that the soup in swan-necked flasks remained sterilized. The bacteria carried by air remained close to the flask's mouth and never completed the long journey down the neck all the way to the soup. Flasks sealed with cotton also remained sterilized. Pasteur's conclusion was clear: there is no such thing as a life force hidden in air. Life comes from life.

Ironically, as we move into modern times and examine viable mechanisms to explain the origin of life, spontaneous generation gets a new boost. Not, of course, from invisible and mysterious powers hidden in the air, but from the bottom-up chemical synthesis of organic compounds from inorganic building blocks. The term *abiogenesis,* as this phenomenon is currently called, is more appropriate than spontaneous generation, which rings of the magical and unexplainable. After all, whatever the details of the chemical processes that led to the first living entity, they were the result of a gradual buildup in complexity from the inanimate to the animated. Unless a ready-to-go living creature sprouted supernaturally out of nowhere—not a very scientific proposition—life could *only* have originated from nonlife. Though the matter isn't settled, Pasteur himself found what many consider a fundamental clue: life is only possible if built from asymmetric pieces.

44 *L'UNIVERS EST DISSYMÉTRIQUE!*

In 1849, well before he won the award from the Paris Academy for solving the spontaneous generation puzzle, twenty-six-year-old Louis Pasteur was working toward his Ph.D. at the École Normale Supérieure in Paris, eager to make his mark among French chemists. He had recently married and needed to firm up his scientific career.

His studies concerned the properties of tartaric acid, a crystalline organic acid present in unripe grapes. Tartaric acid can also be produced in the laboratory through chemical synthesis. Pasteur knew that the acid extracted from grapes and the one produced in the lab had different optical properties, that is, they interacted differently with light. Hidden in this pedestrian fact is a remarkable property of life, perhaps the key to life itself.

But first, a quick briefing on polarized light. As we mentioned in Part III, light is a wave characterized by oscillating electric and magnetic fields. It is a *transverse wave*, in that the two fields oscillate along the plane perpendicular to the direction the wave propagates. For example, imagine that the fields oscillate in the plane defined by the pages of this book. In this case, light would be spiraling outward, toward you. The electric and magnetic fields have another interesting property: they always oscillate at 90 degrees from one another, like the blades of a fan: ✥. Here, if the electric field oscillates from bottom to top and back, the magnetic field oscillates from left to right and back. In general, the fields can oscillate in any direction on the plane, and even rotate along, like the blades of a fan. A linearly polarized light wave has the electric and magnetic fields confined to a single direction of oscillation, as when the fan is turned off. Within this analogy, "rotation of the polarization direction of light" simply means a turn to the left or to the right of the "blades" by a fixed angle.

In 1815, the French physicist and chemist Jean-Baptiste Biot discovered that when light traveled through liquid solutions made out of a number of naturally occurring organic products, its polarization was affected. In the "fan blades" analogy above, these substances could turn the blades of the fan (light's direction of polarization) either to the left or to the right. Pasteur was well aware of Biot's studies. As he wrote in a set of lecture notes from 1860, "[Biot] quite definitely concluded that the action produced by the organic bodies was a molecular one, peculiar to their ultimate particles and depending on their individual constitution."[11] The "action" Pasteur referred to was the ability of these natural organic compounds to rotate the polarization direction of light. With remarkable prescience, Biot had conjectured that such property was related to something going on at the molecular level. But what? Pasteur put Biot's conjecture into firm ground, showing that the optical properties of certain organic compounds—the way they interacted with light—resulted from the spatial structure of their individual molecules.

Building upon Biot's research, Pasteur established that when linearly polarized light passed through a solution of tartaric acid synthesized in the lab, nothing happened: the synthetic solution was optically inactive. But when polarized light passed through a solution containing acid extracted from grapes, *and thus a living entity*, its polarization direction changed (that is, the blades of the fan turned a bit). Pasteur realized that since both substances had identical chemical properties, their molecules had the same types of atoms. What then could cause such puzzling asymmetric behavior? Could living and nonliving substances, even if apparently identical, have different properties? He examined the crystals from both substances under a microscope. He noted that whereas the lab-synthesized acid had two kinds of crystals, the acid from grapes had only one. With tremendous patience, he separated samples of both crystals using tweezers. Passing light through two solutions made with each of them, he demonstrated that the different crystals rotated the polarization plane of light in opposite directions:

> I carefully separated the crystals which were [asymmetric] to the right from those [asymmetric] to the left, and examined their solutions separately in the polarizing apparatus. I then saw with no less surprise than pleasure that the crystals [asymmetric] to the right deviated the plane of polarization to the right, and that those [asymmetric] to the left deviated it to the left; and when I took an equal weight of each of the two kinds of crystals, the mixed solution was indifferent towards the light in consequence of the neutralization of the two equal and opposite individual deviations.[12]

When Pasteur showed his results to Biot, the old man was visibly moved: "My dear child, I have loved science so much throughout my life that this makes my heart throb." Biot's conjecture that the different optical behavior was of molecular origin was vindicated.

Pasteur's findings were indeed dramatic, especially for 1849, when the existence of atoms was not universally accepted. He correctly proposed that the different optical properties of the two solutions of tartaric acid were due to the asymmetrical spatial configurations of their molecules. Tartaric acid molecules can exist in two types, which can rotate the polarization of light to the left or to the right. As Pasteur remarked, "we get identical but not superposable molecules; products that resemble each other like the right and left hands." Pasteur went on to suggest that there are two kinds of molecules in Nature: those that, like water, occur only in one spatial conformation, and those that, like tartaric acid, can occur in two, such that one is the mirror image of the other.

Pasteur's remarkable finding was that the naturally occurring compound only appears in one of its two possible forms while the synthetic one appears in both. Was life selecting a specific molecular orientation?

Continuing with his investigation, Pasteur showed that many organic compounds extracted from living organisms had the same biased optical properties. In one experiment, he added mold to a synthetic sample of tartaric acid. Initially there was no optical activity, as expected. But as the mold grew, so did the optical activity of the sample. Furthermore, the increasing rotation was in the same direction of the naturally occurring acid. There was only one possible conclusion: life has a molecular bias! As Pasteur later wrote,

"The Universe is dissymmetric and I am persuaded that life, as it is known to us, is a direct result of the asymmetry of the Universe or of its indirect consequences." *L'Univers est dissymétrique!*

What prophetic words! The left-right asymmetry of certain organic molecules became known as *chirality*, just as with the rotational asymmetry of neutrinos explored in Part III. As with our hands, the two forms of "chiral" molecules cannot be superimposed one atop the other, being mirror images of each other. The young Pasteur had discovered an amazing property of life. We now know that essentially all amino acids belonging to proteins are left-handed (or, more technically, levorotatory, rotating the plane of polarized light to the left) while all the sugars belonging to RNA and DNA are right-handed (or, more technically, dextrorotatory). Life is indeed asymmetric. The challenge Pasteur left us with is to understand why.[13]

45 THE CHIRALITY OF LIFE

Think again of proteins as long chains of amino acids, pearl necklaces where each pearl is a molecular building block. Imagine that a left-handed amino acid is a white pearl and a right-handed amino acid a black pearl. Life has a clear preference for necklaces with the same color pearls: the crucial molecules for life, proteins, are built from asymmetric backbones. The same is true for the sugar backbones of RNA and DNA. However, in this case the bias goes the opposite way: the sugars are right-handed. It is hard to avoid the suspicion that this molecular bias is somehow related to the origin of life itself. Pasteur was the first to speculate as such:

> Why even right or left substances at all? Why not simply nonasymmetric substances; substances of the order of inorganic nature? There are evidently causes for these curious manifestations of the play of molecular forces. . . . Is it not necessary and sufficient to admit that at the moment of the elaboration of the primary principles in the vegetable organism, an asymmetric force is present?[14]

As with the matter-antimatter asymmetry, we want to understand the causes behind this fundamental imbalance of Nature. At what point in the early evolution of life was the specific chirality of amino acids and sugars chosen? Did it happen right at the start, as simple molecules—probably amino acids—started to interact in the prebiotic soup? Or is the chirality of biomolecules an aftereffect of life, after reproduction had started? Let's consider the two possibilities.

There are two conflicting schools of thought. Some scientists, myself included, claim that chirality came first, that it is hard to imagine how molecular interactions leading to anything like life could have started when both left- and right-handed molecules were present. According to this hypothesis, if initially there were as many left- as right-handed molecular building blocks, as is the case with the amino acids synthesized in Miller-Urey-type experiments, some mechanism must have greatly amplified the concentration of one of the two types until near chiral purity was achieved: back to the pearl necklace analogy, starting with approximately equal numbers of black and white pearls, somehow only chains of white pearls survived, so that building blocks of life have the same handedness. Once chiral purity was reached, reactions followed the pathways toward life.

Alternatively, some scientists propose that it is possible for molecules that are not chiral, that is, that have no distinct left- and right-handed spatial conformations, to have initiated the chemical processes leading to first life. Although there are possible precursors of RNA that are achiral (for example, compounds called peptide nucleic acids or PNAs), I find this possibility quite implausible. The handedness of life is inextricably linked to its molecular functionality. Pasteur agreed: "Molecular asymmetry exhibits itself henceforth as a property capable by itself . . . of modifying chemical affinities."[15] In other words, the shapes of molecules affect how they react with each other. From the perspective of natural selection, handedness was a beneficial attribute, as it facilitated the interactions between complex molecules and, quite possibly, led to the ability to reproduce: handedness and reproduction are deeply related.

Think of the reactions controlling life as a long sequence of lock-and-key mechanisms that can only be activated if the right keys fit in the right order. This is how biochemists interpret the actions of, for example, enzymes in most cellular metabolic and replicating processes.* The same way that jigsaw puzzles only work if the pieces fit right, the reactions that led to the increased molecular complexity of living systems needed spatial specificity. Experiments indicate

* This "lock-and-key" model, although suggestive, is somewhat simplified. The enzymes are not as rigid as a lock suggests, but instead are able to distort slightly in response to the specific demands of the incoming molecules. This variant of the lock-and-key model is called the "induced fit" model.

that even a tiny number of compounds with the wrong chirality truncates the polymerization process, that is, stops the molecular chains from growing. Furthermore, we don't see any hints of a primordial molecular symmetry buried deep in the structures of proteins and nucleic acids: their chiral asymmetry is manifest through and through, all the way to their basic building blocks. In other words, handedness doesn't appear to be a late change of plans in the evolutionary process. From an engineering perspective, it's hard to see why building large molecules with achiral or with equal numbers of left- and right-handed building blocks and then rebuilding them to have only one of the two types would work any better than simply starting with either left- or right-handed blocks from the beginning. So, although there is no conclusive argument either way, I do subscribe to the view that life needs chirally pure, asymmetric initial conditions. Were such conditions prevalent in prebiotic Earth?

In 1953, the same year that James Watson and Francis Crick revealed the double-helix structure of the DNA molecule and Miller ran his spark-of-life experiments, Sir Frederick Charles Frank, a theoretical physicist working at the University of Bristol, in England, published a seminal paper. In it, Frank laid down the three necessary conditions for a solution initially containing a *near equal* number of left- and right-handed molecules to evolve toward chiral purity, that is, to eventually contain either mostly left- or mostly right-handed molecules. First, the chemical reactions had to be such that the more of a given compound is made, the more of it can be made. Such systems are called *autocatalytic*. Some readers may remember Disney's version of Paul Dukas's *Sorcerer's Apprentice* in *Fantasia*. Poor Mickey Mouse is the apprentice who, during his master's nap, steals his magical hat to practice the black arts. Mickey is supposed to fill his master's huge bathtub, bringing water from a well. Mickey bewitches a broom and orders it to do his job. After watching the broom fetch pails of water for a while, he falls asleep, dreaming of controlling the stars and planets with his powers. To his horror, he wakes up to find out that the broom has kept at its work and the floors are beginning to flood. Unable to stop the broom with his clumsy magic, Mickey picks up an axe and chops the broom into pieces. Alas, each little splinter becomes a full-grown broom and keeps at the task. The more he chops the more pieces turn to brooms that flood the castle's floors. So it is with autocatalytic reactions: as more molecules of a compound are produced, they too start to react and more of them are produced.*

The second condition in Frank's model is that there should be a small initial excess of one type of chiral compound, either the left- or the right-handed. In other words, even at the beginning, the symmetry is not exact. (I will explain why soon.) The autocatalytic nature of the reaction then amplifies this small initial excess to a large value. The third condition requires that, when left- and right-handed building blocks combine, that is, when white and black pearls mix, they form molecular chains that are chemically inert, a kind of prebiotic sludge. Frank called this property "mutual antagonism."

During the past few years, different groups in Japan, the United Kingdom, Sweden, and Spain, and my own in the United States, started to study under what conditions Frank's simple model could be made to work in realistic scenarios. Of course, we don't know what chemicals were present on Earth around 4 billion years ago, the ingredients of the "prebiotic soup." Nor do we know much of the atmospheric and environmental conditions prevalent at the time, although it seems that even with a Sun that was 30 percent dimmer there was enough CO_2 in the atmosphere to keep things fairly toasty. But the beauty of modeling is that you can start with fairly general assumptions and obtain results that can, at least in principle, be tested in the laboratory. I stress the "in principle" here. Autocatalytic reactions capable of amplifying a small initial chiral bias are notoriously difficult to achieve in the test tube. In 1995, Kenso Soai's group in Japan obtained the single example we have of an autocatalytic, chirality-biasing reaction. Although it's highly unlikely that the ingredients used were present in early Earth, Soai's beautiful results serve as an important viability study.

More recently, Raphaël Plasson, now at Nordita, the Nordic Institute for Theoretical Physics, in Stockholm, and collaborators pro-

* For readers unfamiliar with *Fantasia,* here is another example: animals that can reproduce without serious predation, such as people or mice in a cage, constitute an autocatalytic system. The more animals there are, the more can reproduce.

posed an alternative model that does away with the explicit—and difficult to achieve—autocatalytic component. So, intriguingly, it's possible to mimic the amplification effects of an autocatalytic reaction network even without one.[16]

46

FROM SO ASYMMETRIC A BEGINNING . . .

Can we construct a viable scenario for how initially symmetric conditions turned asymmetric and eventually led to life? I'd say yes. The steps are clear. To start with, we need both left- and right-handed building blocks. This seems reasonable, since Miller-Urey spark-of-life experiments start with simple chemistry to produce amino acids of both forms in near equal numbers. Also, as Frank suggested, there is need for a small excess of one type of chiral molecule over the other, a tiny initial bias. According to Frank, this small asymmetry is deeply enmeshed with the origin of life itself.

What mechanism or mechanisms could cause this initial imbalance? In Part III we argued that the structures we observe in the Universe, from galaxies to living creatures, ultimately result from a small excess of particles of matter over those of antimatter: for every billion particles of antimatter, a billion-and-one particles of matter are needed. To generate this excess, Sakharov proposed three conditions, which included violations of some of the fundamental symmetries of particle physics. As matter and life needed an initial imbalance, could we follow Sakharov's lead and come up with conditions that dictate how a chiral bias emerged in prebiotic times? Unfortunately, with the origins of life, we don't have such well-defined conditions. In a sense, prebiotic chemistry is not as "clean" as particle physics. Still, there are ways in which a small excess of one-handed molecules could have appeared. Perhaps simple heat could help? Yes, thermal fluctuations alone cause the number of molecules of either type to vary by small amounts in different places.[17] Remarkably, using either autocatalytic reactions or activated amino acid models discussed previously, it's hard but not impossible to amplify such a small excess of one type of molecule into the overwhelming majority in a relatively short time. As Frank remarked in his seminal paper, such a mechanism—if it could actually be demonstrated in the laboratory and not only in theoretical models—could generate regions (for example, shallow ponds) with molecules of opposite handedness coexisting side by side and fighting for dominance: natural selection at the prebiotic level. Still, once more detailed calculations are done, it appears that thermal fluctuations alone may be too feeble to promote a large enough initial bias. We need something better. Pasteur had anticipated this as well:

> Do these asymmetric actions, possibly placed under cosmic influences, reside in light, in electricity, in magnetism, or in heat? Can they be related to the motion of the earth, or to electric currents by which physicists explain the terrestrial magnetic poles? It is not even possible at the present time to express the slightest conjecture in this direction.[18]

But now, more than a century and a half after Pasteur wrote these lines, we *can* conjecture.

Over the past decades, many scientists have proposed ways for generating an initial excess of molecules of a specific handedness, a chiral bias. The best-known mechanism relies on the spatial asymmetry of the weak nuclear interactions, the parity violation we discussed in Part III. The reader may recall that neutrinos only appear in their "left-handed" form. If Nature already has a bias at the nuclear level, could it trigger the molecular bias we see in living creatures? This would be, by far, the most satisfying solution, a beautiful link between a fundamental asymmetry of matter and that of life. If true, it would also have a startling consequence: *any* extraterrestrial life form, irrespective of where it is in the Universe, would have left-handed amino acids (and, presumably, right-handed sugars). Life, like neutrinos, would carry the same universal chiral fingerprint.

In spite of many worthy efforts by colleagues, it's unlikely that the weak interactions—which are active only inside the atomic nucleus and hence at distances *much* smaller than a molecule—could be responsible for life's asymmetry. There is one main reason: the biasing effect is extremely small. Biomolecules are immense structures when compared to atomic nuclei. It's hard to see how amplification mechanisms could be this effective, even in highly unstable systems. The energies involved in the chiral bias from weak interactions are thousands of trillions of times smaller than typical bonding energies in sugars. As we have seen earlier, the simplest solution, even if the most appealing, may not always be the right one. Furthermore, we would still have to understand why the chiral biasing would go in opposite directions for amino acids and sugars.[19]

Could, as Pasteur had suggested, some kind of radiation have caused the biasing of the prebiotic soup? Imagine that during its infancy, the solar system passed near a star-forming region. Such regions are strong emitters of circularly polarized UV radiation. Many believe that this kind of radiation could have caused an initial chiral bias that was then amplified and eventually locked into the biomolecules of life. A consequence of this mechanism is that all of the solar system would have the same bias: if we could find chiral molecules in, say, Mars or in in Saturn's moon Titan, they would share the same bias, such as having predominantly left-handed amino acids. Outside the solar system, however, different conditions could push the biasing in the opposite way. Unlike the weak interactions, biasing from radiation would not be universal. Again, this scenario has difficulties. Not only is it hard to find compelling candidates for star-forming regions in our neighborhood of 4 billion years ago, but the efficiency of the chiral biasing under UV light in space is still under active debate.

A third way of generating an initial excess of molecules of one handedness is related to the "where" question. If the reactions leading to the first biomolecules occurred in mineral or clay surfaces, they could have been induced to take sides: the crystal structure of the surface could have acted as a template, a sort of chemical railway track fixing a single spatial orientation for the molecules.

The above scenarios offer a measure of how much effort has been dedicated to solving the handedness-of-life mystery. There is, however, one key aspect that has been largely overlooked and that, in my opinion, could be decisive. It is what got me excited about this research topic in the first place. During the summer of 2006, I noticed the many similarities between the origin of life's handedness and the origin of the matter excess: cosmology meets biology. Just as the matter excess was cooked in the unstable environment of the early Universe, the prebiotic chemical soup was cooked in the unstable environment of the early Earth. Thus it is reasonable to suppose that what was happening on our young planet had a crucial influence on the emergence of first life, including the choice of molecular handedness. To what extent did the young Earth's active environment influence the origin of life? Pasteur hinted that this influence would be crucial, as did Frank one hundred years later. In 2005, Axel Brandenburg and Tuomas Multamäki of Nordita argued that turbulence could have sped up the amplification mechanism toward one handedness. Meanwhile, Dilip Kondepudi of Wake Forest University and his collaborators and Cristóbal Viedma from the University of Madrid observed that stirring a solution containing both left- and right-handed chiral crystals accelerates the bias toward a single handedness, without any obvious preference of one over the other. The solution could end up either with left- or right-handed molecules. But it would be chirally pure.

Since we don't have a time machine to go back to early Earth, how can we study the impact of environmental events on the choice of life's chirality? The answer involves lots of very fast computers. We can use them as laboratories, in effect creating a model Earth where chemicals react in an unstable environment. At Dartmouth, Joel Thorarinson, Sara Walker, and I set up to model how environmental effects could have influenced the reactions in Frank-like systems. Chemical reactions are very sensitive to temperature fluctuations and to changes in the concentrations of the reactants. If we go back to primeval Earth, we can imagine that external disturbances—falling meteorites, volcanic eruptions, rampant earthquakes—greatly influenced the prebiotic reactions at different places. The results of our research were quite dramatic: environmental effects, if strong enough, could have completely erased a previous excess of either type of handedness. Put another way, if chemical reactions are nicely evolving toward, say, a clear excess of left-handed amino acids, an environmental disturbance could flip the bias toward a right-handed excess.

One way of visualizing this is to imagine hundreds of coins placed on a stretched rubber sheet. Each coin represents a molecule, with heads being one handedness (say, left) and tails the other. The rubber sheet can vibrate with varying degrees of violence in different spots. Think of small environmental disturbances as small-amplitude vibrations of the rubber sheet and large ones as large amplitude vibrations. Imagine further that all coins are initially placed on the sheet with heads facing up: a "chirally pure" initial condition. Vibrations will make the coins jump about. Although small vibrations won't be able to flip heads into tails, large ones can. Above a threshold amplitude for the vibrations, the system will become so agitated that the coins will have a fifty-fifty chance of being flipped. If the violent vibrations are turned off, we should expect to see approximately half the coins as heads and half as tails: the initial arrangement with all coins as heads is destroyed. We say that the system passed through a "critical point." Past this point, the initial ordered arrangement (all coins as heads) is destroyed. If you substitute coins for molecules and heads and tails for left- and right-handedness, you see how environmental effects can erase any previous bias toward a single handedness. Essentially, after a violent external event, the system is reinitialized, with approximately equal numbers of left- and right-handed molecules. Then, after things calm down, the reactions run again, amplifying one or the other handedness.

We can use our computer simulations of early Earth to find the critical intensity of environmental effects needed to flip the handedness of molecules in a small "virtual pond." Our results indicate that such events were quite probable (with probabilities above 60 percent in many instances), given reasonable assumptions. As a consequence, any initial excess of a given handedness could have been flipped around several times in the course of Earth's history. In other words, each violent event had the power to erase any previous bias toward a specific handedness: there is no memory of the past. Our knowledge of Earth's prebiotic past can never be complete. After some time, somewhere around 3.8 to 3.5 billion years ago, environmental disturbances became sufficiently weak and prebiotic chemistry sufficiently robust so that a given handedness prevailed. According to our model, the fact that the proteins of terrestrial creatures are built out of left-handed amino acids is a fluke; amino acids could have turned out to be right-handed.

Our model makes another strong prediction: extraterrestrial life can have *any* handedness. So, unlike the biasing mechanisms from weak interactions or from UV light, if we could find amino acids elsewhere in the solar system they could be either right- or left-handed, with no obvious preference. Now, extraterrestrial amino acids *have* been found in some meteorites, most notably in the Murchison meteorite mentioned before. Jim Cronin, Sandra Pizzarello, and others have found a small excess of a few left-handed amino acids (for example, up to 15.2 percent in samples of isovaline). This could be interpreted as evidence that our predictions are wrong and that an initial bias acted on the solar system as a whole, as other mechanisms predict. However, that would be premature. For one thing, the excess toward left-handed amino acids is not complete as it is on Earth and is only mostly for isovaline. Other amino acids show a much smaller bias. For another, our results are statistical and need many samples to be corroborated. At present, all we have is a small excess of some left-handed amino acids found in the Murchison and the Murray meteorites.* Although very important, this evidence cannot be used to state that amino acids in the solar system are predominantly left-handed. Apart from the excess being small and affecting only a few of many amino acids, there is the danger of contamination, that is, the idea that terrestrial chemical processes biased the amino acids in the meteorites, although much care is taken to avoid this. Also, and very importantly, these amino acids were not extracted from a living creature. It is quite possible that the chiral bias is related to the chemistry of life and not to whatever chemistry took place on a rock traveling across the solar system for 4 billion years.

How can we determine if our predictions concerning the arbitrariness of Earthly life chirality are correct? If more evidence is found across the solar system indicating that amino acids are indeed overwhelmingly left-handed, then our hypothesis must be abandoned and we must accept that there was some biasing mechanism that acted on the whole of the solar system, possibly the Universe, and that somehow local environmental effects were never powerful enough to alter the bias. I remain skeptical. Even if we are wrong, we will learn something about the preconditions for life. This is how science advances; sometimes you win, sometimes you lose. Only data can decide. We only know what we can measure.

* The Murray meteorite fell on Calloway County, Kentucky, in 1950.

Our model reminds me of a dramatic twist to the theory of evolution proposed in the early 1970s, the *punctuated equilibrium* hypothesis of Niles Eldredge and Stephen Jay Gould. Inspired by the work, we called our paper "Punctuated Chirality." Contrary to the commonly accepted gradual evolution of the species, Eldredge and Gould's punctuated equilibrium scenario suggested that life evolves in fits and starts; periods of relative calm, when not many new species appear, give way to periods of accelerated speciation. The causes for these changes are generally attributed to natural cataclysms, such as meteoritic impacts and volcanic eruptions. A famous example is the collision with a ten-kilometer-wide asteroid that ended (or greatly contributed to ending) the dinosaurs and about 40 percent of life on Earth 65 million years ago. The impact marks the boundary between the Cretaceous and Tertiary periods, when mammals began to dominate the scene.[20] In a way, we are extending the punctuated equilibrium hypothesis of Eldredge and Gould to prebiotic times, proposing that environmental events played a key role in determining the chirality of terrestrial life. If we are right, the deep relationship between the history of Earth and the history of life started even *before* life.

47 WE ARE ALL MUTANTS

Readers who, like me, love old horror movies must have witnessed many metamorphoses of man into werewolf on the silver screen. My favorite is the 1941 classic *The Wolf Man:* under the full Moon, a lugubrious Lon Chaney, Jr., starts to grow hair and fangs as gentle human turns to human-killing monster.

Even though the tremendous feats of makeup and special-effects artists of the '40s look almost comical compared to the computer-aided graphics of today's movies, and now teenagers laugh when they watch old horror classics with flattering comments like "Dad, this is ridiculous! Were you actually scared of this junk?" one of my criteria for a movie's quality was the realism of the man-to-beast-to-man metamorphosis. The scarier movies were the ones where changes from human to beast and back to human were the most continuous, the most gradual.

Maybe some readers are familiar with the old "Gorilla Woman" carnival act, featured in the James Bond movie *Diamonds Are Forever* and one of the Marvel comic "Freaks," where a beautiful girl in a scant bikini (at least in Brazil) gradually turns into a furious, growling hairy ape in front of your very eyes.[21] The scariest of these acts were the most realistic, where the metamorphosis occurred gradually. In fiction, to pass as truth the lie must be well told.

Even before the days of the pioneer geologist Charles Lyell in the 1830s, gradual transformation, also called uniformitarianism, had been a leading idea in geology. The Earth changes slowly and gradually over enormous periods of time, much longer than humans can contemplate. To find the record of these changes we must resort to rocks. They are our links to Earth's distant past.

When the young Charles Darwin set off to explore the world

aboard the HMS *Beagle*, Lyell's book was at his bedside table. As his thoughts about evolution developed, Darwin realized that the history of life on Earth and the history of Earth itself, its geological history, were deeply connected. While geologists looked at rocks to reconstruct the history of Earth's past, paleontologists should look at the fossil record for evidence of the gradual transition from one species into another. These gradual transitions would be the signature of natural selection in action. However, as Darwin acknowledged in his *On the Origin of Species*, things were not so simple. The fossil record provided no evidence of "infinitely numerous transitional links." He searched for guidance in geology, arguing that there were gaps there as well: "Why then is not every geological formation and every stratum full of such intermediate links? Geology assuredly does not reveal any such finely graduated organic chain." Darwin then masterfully invokes the link between geology and life to preempt this kind of criticism to his theory: "The geological record is extremely imperfect and this fact will to a large extent explain why we do not find interminable varieties, connecting together all the extinct and existing forms of life by the finest graduated steps. He who rejects these views on the nature of the geological record, will rightly reject my whole theory." In other words, given that we can't find all the intermediate rocks to illustrate the gradual transitions from one geological era to another, we should not expect to find all the intermediate life forms between extinct and modern living creatures. Even so, Darwin hoped that paleontologists would try as hard as possible to fill in the gaps, hunting for the "intermediate links."

Darwin's approach to evolution mirrors the metamorphoses of the old horror movies: the best man-to-werewolf sequences had the fewest breaks in the sequence, gradual transformations with as many intermediate steps as possible. Darwin believed that these "finely graduated steps" would have occurred, for example, as dinosaurs transitioned into birds (as in the famous archaeopteryx fossil) or, like the end of the Gorilla Woman act, as apes morphed into humans. Gradual transformations would accumulate and reach a point beyond which it would no longer be possible for changed members of a population to breed with their unchanged peers: a new species would be born. Since the evolutionary road from species A to species B was believed to have been nice and smooth, the fossil record ideally should reflect these gradual changes; in reality, however, the fossil record would necessarily have gaps.[22]

Opposing the gradualist approach, the catastrophist school would claim that cataclysmic events greatly determined the history of Earth and, consequently, of life on Earth. A large falling meteorite, a sequence of violent volcanic eruptions, abrupt climate changes, earthquakes and tsunamis, all of these could have disrupted the stately pace of life's evolution either locally or even globally in time scales much shorter than traditional geology would allow. Decades of lively debate and much research and data gathering led to a consensus integrating both views: Earth's geological history is a slow, gradual process marked occasionally by violent and disruptive events. The impact of such events on life's early history has only begun to be understood.

How can we distinguish between the two pictures? It's impossible to demonstrate that gradualism works if we accept that the fossil record is incomplete. In other words, strict gradualism is not falsifiable: unless we invent a time machine capable of traveling to the past so as to observe the details of speciation firsthand—not something even modern physics allows—we will *never* be able to argue that gradualism is indeed correct. We only know what we can measure. This problem must have haunted Darwin and his followers. On the other hand, if a combination of catastrophism and gradualism is accepted, then we should expect discontinuities in the fossil record. The most famous example, of course, is the abrupt extinction of the dinosaurs across the boundary of the Cretaceous and Tertiary eras.

The key ingredient missing in Darwin's theory of evolution was the specific mechanism for how species change. He didn't know about genes and mutations, DNA duplication, or meiosis. The crucial addition of genetics to Darwinian evolution is sometimes called the "Modern Synthesis." Every living creature has its own genetic code, or *genotype*, detailing the molecular expression of its physical characteristics, or *phenotype*. We can think of it as a recipe to construct a specific living creature in detail. During reproduction, this set of instructions builds offspring through a series of complex molecular interactions. When the building is 100 percent efficient, the offspring inherits the parents' genes. If reproduction is asexual, the daughter cells will carry the same genetic information as the parent cell. This is what happens, for example, in ordinary tissue growth, when, say, a liver cell splits into two identical ones. Things would be quite bad if a liver cell gave rise to heart cells as it divided. In sexual reproduction, there is a mixing of genes from father and mother. In both cases, if the building process is defective, the daughter cells will have gene sequences that differ, even if only slightly, from those of the original sources. The imperfect results of the genetic offspring-building process are called mutations.[23]

As life evolved on Earth, the genetic copying process no doubt became increasingly more efficient and sophisticated. Duplication in the first life forms was probably prone to many errors. Presumably there were more mutants than species proper. As time passed and life persisted, the reproductive apparatus evolved. Creatures grew more adapted to their environments and thus more diverse from others in different environments. A kind of microbial speciation had begun. As Darwin noted when he visited the Galapagos Islands, geographic isolation greatly accelerates speciation. Although the complex cells of modern multicellular organisms are a far cry from the simple single-celled creatures that lived 3 billion years ago, they do have much in common. Our single-celled predecessors must have developed the genetic language that we identify today in all living creatures based on four nucleotide bases and on proteins built out of twenty amino acids. Imagine, then, that some 3 billion years ago, after much trial and error, a single-celled organism achieved reproduction as we now observe it, via DNA duplication. That ancient creature would be the last universal common ancestor (sometimes called LUCA), the ancestor of all currently living animals, the root to the tree of life. We can appreciate now the true meaning of Darwin's lyrical (and prophetic) words, "From so simple a beginning endless forms most beautiful and most wonderful have been, and are being, evolved." What Darwin didn't yet know is that the remarkable branching out that led from LUCA to humans was due to mutations.

A useful way to think of DNA is to picture a zipper, with its two sides connected by teeth. There are only four kinds of teeth, the nucleobases, and they have a very specific pair matching: adenine (A) pairs only with thymine (T), while guanine (G) pairs only with cytosine (C). So, if on one side of the zipper there is a sequence like

A-C-A-T-G, on the other side the sequence *must* be T-G-T-A-C. A gene is simply a specific sequence of bases that contains the instructions for a specific function. Different genes do different things. For example, they may contain instructions for making different proteins, such as hemoglobin. The complete set of genes of a living creature constitutes its *genome.*

Now that we have introduced all these concepts we can explore what a mutation is at the genetic level. During reproduction, the DNA molecule unzips itself and two new copies are made, one from each side. For this to happen, genes must be read off in the right order. Think of a gene as a sentence made of a few words. If a typist copying the sentence switches a few letters, the sentence may lose its meaning. Likewise, if there is an error in the gene sequencing, the offspring will carry an imperfection: a mutant is born. In the majority of cases, mutations are neutral and have no obvious effect on the survival of the animal. The ones that do have an effect, that is, mutations that play a role in natural selection, are often deadly or severely crippling to the creature. Typos don't usually improve a sentence. For example, a mutated lung cell may lead to the growth of an abnormal cell patch causing cancer. However, in rare events a mutation may actually be beneficial, giving the creature an edge in the survival game. The cliché example is that of giraffes. If most low-hanging leaves have been chewed away, the animal with the longest neck will be the one able to reach leaves at higher branches. This mutant will eat more, become stronger, and mate more effectively. Soon a growing number of his descendants will be carrying the long-neck gene and the giraffe population will slowly change from short-necked to long-necked. Mutations are the main source of genetic variation, the main engine behind the remarkable diversity of life. Had reproduction been perfect, species would not have undergone mutations and survived the many long- and short-term environmental changes they have faced over time. In other words, without mutations the tree of life wouldn't have branched and life would have been a failed experiment.

We owe our existence to the imperfections of genetic reproduction. As made clear by the great evolutionary biologist Ernst Mayr, we can see why a species that was geographically isolated from its ancestors will change in time. Adaptive mutations will remain geographically confined and slowly (or not so slowly initially) transform the species as a whole. If, after some time, the geographic blockage disappears or is somehow overcome (for example, the glacier melts due to global or local warming, or erosion carves a passage through the mountain peaks, or the animals can withstand longer migrations) the mutant species may return to its ancestral home. Millions of years later, a paleontologist may uncover a fossil record containing both specimens, without any gradual changes between them.

Even though as billions of years passed (and thanks to mutations) the tree of life branched into many different directions, we all carry the roots of our origins in our genes, all the way back to the last universal common ancestor. The history of life on Earth is a powerful illustration of how imperfections give rise to creation and diversity. From the chiral backbones of proteins and DNA in the prebiotic soup to the endless forms most beautiful of today's Earth, we exist due to the small but essential inaccuracies of the genetic reproduction. If we could rerun the course of history, as gods toying with fashioning worlds, life on Earth would certainly have taken a different turn.

Mutations occur randomly, either at the level of DNA replication or prompted by external sources such as ultraviolet and X-ray radiation. *Our existence owes itself to a very specific set of mutations, unleashed under a very specific set of environmental conditions.* Change the mutation or the environment and the branching of the tree of life also changes. To put it more dramatically, within a different sequence of events we would probably not exist. We are the products of a very specific chain of events. If no asteroid had eliminated the dinosaurs 65 million years ago, who knows if mammals would have prevailed. Contrary to the belief of many, natural selection does not prescribe that intelligent species will necessarily develop given enough time. Although certain evolutionary traits, such as eyes, may reappear under various circumstances—a phenomenon some biologists call "evolutionary convergence"—this simply reflects their usefulness as an adaptive device. In the case of eyes, they clearly make life much easier under a star that radiates mainly around the yellow portion of the electromagnetic spectrum. You need to see to survive on Earth but you don't need to be smart. There is no preordained end goal to natural selection, only efficient adaptation to a specific environment. The dinosaurs reigned supreme for more than 100 million years and were, as far as we can tell, quite stupid.

Viewed this way, the fact that we exist at all, that the marriage of a very specific planetary history with a very specific sequence of random mutations led to an intelligent species capable of pondering its origins and whereabouts, is nothing short of wondrous. Here we have two choices: either we ascribe our existence to a miracle perpetrated by supernatural powers, or we accept how fragile we are, how fragile life is.

Modern science has shown how the chain of life started even before stars were born, when matter ruled over antimatter in the primordial cosmos. The seeds of life, the chemical elements that make up all proteins and DNA molecules, were forged—and continue to be—in the hearts of dying stars, trillions upon trillions of them, belonging to the hundreds of billions of galaxies floating in the vastness of space. Around many, probably most, of these stars, planets and their moons move in their stately choreography. How humbling it is to think of the countless other worlds spread across the universe. And how hard it is not to wonder if life has evolved elsewhere and, if it has, to what level of complexity . . . Even though we have not yet made contact with extraterrestrial life, the possibilities are tantalizing, and to some, terrifying. The alternative, that we are condemned to cosmic solitude, alone to ponder questions of life and death, would imply that we have only ourselves to rely upon, that we are truly precious and rare. Are we ready to fulfill our cosmic role? To contemplate the existence of extraterrestrial intelligence is to put humanity in front of a mirror. In the final part of this book we turn to these questions informed by what we've learned about life, the cosmos, and the limits on our search for knowledge.

PART V

The Asymmetry of Existence

48

FEAR OF DARKNESS II

An old man sits in the dark, thinking of his dead lovers. Night has fallen fast, faster every day . . . Looking out the window, he sees a star blinking in the distance. "Is that star even there, shining still?" He can't know. No one can. "How many worlds are out there, invisible to us, too far to see, too dim? We see so little; we know so little."

He thinks again of his dead lovers. "Do they still love me up there in heaven? Or are they stardust now, sprinkled across space? What will I become? An angel? Dust?" All his life, he dreamed of certainties where none was to be had. First, he hoped to fill loss with faith. Then, he tried with knowledge. He searched everywhere for meaning and found out that there were no simple answers, no final, cohesive picture; no grand plan for Creation. He struggled, not wanting to accept that there were limits to knowledge, that he could never know everything. He felt small and incompetent. "If I can't make sense of the world, who am I?" For a long time, he refused to give up. He couldn't embrace the simplicity of not knowing. But slowly, things started to change. What was heaviness became lightness. What seemed lost became a new path. To not be a part of a grand plan was liberating. He could always ask questions and keep learning about himself and the world. He could always love and hope to be loved. To be alive and to be remembered, he finally understood, was what mattered. Beyond that there is only darkness.

49 IS THE UNIVERSE CONSCIOUS?

"All philosophy is based on two things only: curiosity and poor eyesight . . . the trouble is, we want to know more than we can see." Thus begins the philosopher's soliloquy in Bernard le Bovier de Fontenelle's superb *Conversations on the Plurality of Worlds*, published in 1686, a year before Isaac Newton's *Principia*. We are curious about the world and want to know the answers to the big questions. The trouble is, our tools allow us to see only so far. Undaunted, we let our thoughts and imagination carry us beyond the measurable, into the realm of the unknown. We search for answers with the two means we have at our disposal: faith and reason. Some choose to look at the world mostly through faith, believing the natural order of things to be the result of supernatural interference. Although science may play a role in their lives—they may take antibiotics and may understand that the digital revolution is the product of our understanding of the atom—it plays none when it comes to big questions such as the origin of the Universe, the origin of life, or what happens after we die. We may call members of this group "supernaturalists." Others choose to look at the world through reason alone, believing that natural phenomena are the result of causes firmly rooted in immutable laws. The scientific method is the only viable approach to obtain such laws and hence to establish a rational foundation for Nature. This second group does not call for any supernatural explanations of the unknown. We may call its members "naturalists." However, even here there is room for a combination of reason and faith. We just must be careful to define "faith" as the unproven belief in something. Whether this "something" is supernatural or natural distinguishes the supernaturalists from the naturalists.

Einstein, for one, would claim that neither faith nor reason alone is enough, that they need each other. "Science without religion is lame, religion without science is blind," he famously wrote. However, his faith was not in some unexplainable supernatural causation, but rather in a Platonic ordering of Nature hidden at the heart of existence, fragments of which the human mind has the privilege of occasionally unveiling through the workings of science. As the patriarch of the unification trend in modern theoretical physics, Einstein believed that the order we find in Nature, the mathematically precise laws describing the motions and interactions of material bodies, was an indication of a deeper ordered pattern: "I believe in Spinoza's God who reveals himself in the harmony of all that exists, but not in a God who concerns himself with the fate and actions of human beings."[1] Like Kepler, Einstein longed for the harmonies.

We are meaning-seeking creatures. We strive to understand why the world is the way it is and why we act the way we do. We attribute causes to actions, causes that serve a purpose. We exercise, we work, and we love, trying to live our lives in the best possible way, balancing duty and pleasure. To be confused, to be lost, is usually equated with having no sense of direction, no purpose. No good can come out of it. Should Nature be any different? We look at the world and see order all around. We see the passing of the seasons and how it affects the growing of plants and the behavior of animals; we see the orderly motions of the celestial bodies and the remarkable ability of our mathematical laws to describe so much of the world. Surely, the laws are telling us that Nature has a sense of direction, too? Can we then say that there is purpose in the Universe? That the Universe has a mind of its own and we, somehow, are a consequence of this cosmic script? As the great physicist John Wheeler liked to put it, "How come existence?" In his recent book, *Cosmic Jackpot*, physicist and natural philosopher Paul Davies addresses Wheeler's question, offering a very thorough and compelling review of its many possible answers.[2] I will briefly go over what has been said so we can see where I differ and what this book has to offer in this conversation.

In one popular notion, sometimes called the "Accidental Universe" or the "Absurd Universe," the Universe is an accident, life is an accident, and there is no purpose whatsoever in anything that happens. The fact that we developed the ability to think and ask questions is also a fluke. What interests me here is that many people, in particular Unifiers and, of course, religious groups, interpret this view as a "cop-out." Somehow, by accepting the accidental nature of physical and chemical processes that take place in Nature, we give up the search for deeper relationships among life, mind, and cosmos. The name "Absurd Universe" already attaches a negative value judgment to this view. It shows how hard it is for us to accept that maybe the cosmos is meaningless, that there is no hidden code of Nature—either scientific or God-made—to justify our existence.

The critics of this idea miss the fact that a meaningless cosmos that produced humans (and possibly other intelligences) will never be meaningless to them (or to the other intelligences). To exist in a purposeless Universe is even more meaningful than to exist as the result of some kind of mysterious cosmic plan. Why? Because it elevates the emergence of life and mind to a rare event, as opposed to a ubiquitous and premeditated one. For millennia, we believed that God (or gods) protected us from extinction, that we were chosen to be here and thus safe from ultimate destruction. This kind of comforting thought removes responsibility of our survival from ourselves, conveniently placing it on a supernatural guardian (or guardians). When science proposes that the cosmos has a sense of purpose wherein life is a premeditated outcome of natural events, a similar safety blanket mechanism is at play: if life fails here, it will succeed elsewhere. We don't really need to work to preserve it. To the contrary, I will argue that unless we accept our fragility and cosmic loneliness, we will never act to protect what we have.

The Unifiers, on the other hand, believe in a grand plan, a mathematical hyperstructure that underlies all there is. The plan includes explanations for the origin of the Universe and for all the properties of the fundamental particles of matter. All of Nature derives from this single formulation, called a Theory of Everything. There is a radical and a moderate version of this viewpoint. In the radical view, there is a single theory whence everything follows. In this case, all the properties of the cosmos and of matter are contained in this formulation and there are no free parameters to be chosen: the theory is constructed from rigid mathematical relations that reflect the highest symmetry. In the moderate version, there may be many unified theories, each describing a self-contained reality, for example, as part of a multiverse, and we just happen to exist in the one that explains the Universe we are familiar with. Neither view has anything to say about the emergence of life, instead limiting its focus to the inanimate aspects of reality. However, as David Gross, a Nobel-laureate particle physicist and believer in the radical version of unification (and with no patience for multiverse-like arguments), once told me, life and intelligent life should be ubiquitous in the cosmos. If there is a grand plan for the Universe and we emerge from it, then it follows that we shouldn't be special; similar conditions elsewhere should generate sentient beings. This view has been elevated to a principle known as the "principle of mediocrity," an extension of the Copernican principle that states that Earth is an unexceptional planet and that "we are just one out of a multitude of civilizations scattered throughout the universe."[3] Given what we already know of the history of life on Earth and in the solar system (and some of the other stellar systems astronomers are now observing), it is very hard to agree that Earth is unexceptional, and even harder to believe that there are countless civilizations out there. I will argue that such a belief, apart from ignoring a host of information coming from research in astrobiology, can have very negative philosophical and social implications.

As I described before, in a multiverse model our Universe would be one of many, possibly infinitely many, other universes. In some versions, such as those related to the string landscape, different universes may have altogether different properties. In some the electron may have a different charge or mass; in others it may not exist. Most of the universes would be barren, unable to sustain any kind of life. Our Universe happens to be the one where things work just right for conscious beings to emerge out of the laws of physics and the constraints of biology. Believers both in a unified theory and in the multiverse hope that some sort of selection principle would pick, out of the countless variety of cosmoids, the one corresponding to our Universe. For example, each valley of the string landscape would be a possible universe and only *one* valley would be ours. A serious difficulty with the multiverse idea is that it is very hard to test. Proposals are out there, but it's fair to say that they remain far-fetched. In a framework where everything is possible nothing is truly understood, especially if competing hypotheses cannot be falsified. An extreme example is the ecstatic Platonic version of the multiverse proposed by physicist Max Tegmark, where all imaginable worlds exist in some sort of hypothetical mathematical fantasyland.[4]

There are also those who believe that the existence of life in the Universe is not an accident or a consequence of a unified theory of nature. There is some sort of overarching "life principle," acting beyond the mere physical laws we presently use to describe the natural world. This principle may one day be discovered, although of course we have no clue yet how it could be. The laws of physics and the laws of chemistry as presently understood have nothing to say about the emergence of life. As Paul Davies remarked in *Cosmic Jackpot,* notions of a life principle suffer from being teleologic, explaining life as the end goal, a purposeful cosmic strategy. The human mind, of course, would be the crown jewel of such creative drive. Once again we are the "chosen" ones, a dangerous proposal. Unless supported by scientific arguments, it's very hard to see how different the life principle really is from attributing our existence to an unexplainable divine act. On the other hand, as so many examples in the history of science have shown us, what today may seem fanciful or even supernatural may one day be explained by science. The scientific quest for a principle relating the emergence of life to physical and chemical causes is, in fact, an excellent research topic. The question, of course, would remain as to why, or to what purpose, would the Universe want to create life and mind. Instead of the "Why are we here?" question, we would have the "Why is the universe here?" question. Could the Universe have created us to figure itself out? Arguments shifting the "mind of God" to the "mind of the cosmos" perpetuate our obsession with the notion of Oneness. Our existence *need not* be planned to be meaningful.

In the 1970s, astrophysicist Brandon Carter proposed the so-called *anthropic principle,* according to which intelligent observers are a consequence of specific physical properties built into the fabric of the cosmos. In the strong version of the principle, the Universe has engendered observers, again reflecting a mysterious teleology. As Davies writes, "somehow, the universe has engineered its own self-awareness." (Hence the "mind of God" to "mind of the cosmos" shift.) In a sense, the Universe itself is alive, a creative entity capable of engendering creatures able to reflect upon itself. To me, even though there is no God talk here, a self-aware cosmos capable of creating life and mind is like an omniscient deity, a vain one at that, capable of existing both within and without time. Life and mind are truly important, but, in my view, not for this reason.

In the weak version of the anthropic principle, observers appear only in a statistical sense: within the multiverse hypothesis, there is a passive selection mechanism allowing for life and mind to emerge in a specific subset of cosmoids, including, of course, ours. At a superficial level, the weak anthropic principle states the obvious: we are here because the Universe has the right properties for us to be here. At a deeper level, as pioneered in the work of Steven Weinberg in the mid-1970s and more recently by Alex Vilenkin, Jaume Garriga, Andrei Linde, and other cosmologists, the principle can be used to place reasonable bounds on important cosmological parameters, such as the amount of dark energy in the Universe. Still, it's very hard to avoid feeling that we are not really learning anything very deep or new, even when we succeed in obtaining such bounds.

50 MEANING AND AWE

It's clear that opinions on the subject diverge widely. Yet with the exception of the so-called "absurd universe," they all hinge on the notion of Oneness in one way or another. Either there is a unified theory, or there is an all-encompassing multiverse, or there is a self-aware, life- and mind-creating cosmos. You could even combine all of these and say that there is a self-aware multiverse where unification is manifest. Should that be a surprise? In this book, I have argued that the modern scientific search for a single, overarching explanation of existence is rooted in monotheistic thought, functioning as a rational substitute for the notion of a supernatural God. Unified theories, life principles, and self-aware universes are all expressions of our need to find a connection between who we are and the world we live in. I do not question the extreme importance of understanding the connection between man and the cosmos. But I do question that it has to derive from unifying principles.

Could it be that there is no final truth out there to be discovered, that we are indeed the product of a sequence of accidents? The common view is that if this is the case all is lost: if the Universe is a fluke, we are left without a sense of purpose, without a direction to pursue our search for meaning. I couldn't disagree more. On the contrary, I would argue that it is *precisely* our insistence in searching for "unique" and "final" explanations that is delaying progress in our true search for meaning. The more arcane the search for Oneness, the further removed we become from Nature and from the pressing problems of our time. Even worse, the whole search may become an escape, the "opiate of the mind," to paraphrase Marx's famous maxim. I am reminded once again of Kepler's anguished words, uttered almost as a prayer during a time of personal and political chaos: "When the storm rages and the shipwreck of the state threatens, we can do nothing more noble than to lower the anchor of our peaceful studies into the ground of eternity."[5] It is no wonder that Socrates, tired of the pre-Socratic search for the One, "called down philosophy from the skies," as Cicero wrote. It's time to lower the anchor of our peaceful studies into the ground of reality.

Inherent in our drive to understand is a deep reverence for the beauty and drama of the natural world, the awe we experience at the grandeur of Creation. Contrary to what so many have believed for millennia, there is no a priori connection between this awe and the "longing for the harmonies," the search for a final explanation for all there is. The awe that drives our search for meaning need not be attached to an antiquated notion that all is one. We can explore the ocean without having to look for a mythic treasure. We can explore the mysteries of the natural world without having to believe that they all spring from some Final Truth, the hidden code of Nature. All is many, not one. There are countless manifestations of how change engenders form, how asymmetries create the structures we see in the world around us. At the apex is life in all its staggering diversity, the result of molecular asymmetries and genetic mutations. Can humans learn to celebrate life and mind without elevating either to God-like status? I hope so. To say that life is the creation of a purposeful cosmos is to promote it to a state of pseudoreligious immunity, independent of our actions and choices. That, to me, is a grave mistake, as it frees us from our responsibilities as the only living creatures (that we know of) mindful of life's importance. "Surely, if the cosmos made us it has made other mindful creatures," many would argue. Well, we don't know that and, as I will advance below, we probably won't know it for a very long time, if ever. So it's up to us to do something and to act quickly. Life is awesome precisely because it is rare and fragile, the precious result of a series of accidents. If we don't care for it, chances are the Universe won't, either.

51 BEYOND SYMMETRY AND UNIFICATION

One of the major difficulties in the long-standing feud between science and religion is the accusation directed at scientists that they took God away from people and offered nothing in return. This is what the Brazilian audience member we met earlier in the book accused me of doing when I offered a scientific explanation for the Big Bang. If all is reason and causation, if everything rests on rational explanations, what place for our human emotions, the pain of loss and despair, for our capacity to love?

Perhaps the greatest of all injustices of such an indictment against science is the notion that a naturalist—as opposed to a supernaturalist—description of existence is devoid of magic and wonder. Many distinguished scientists, including Carl Sagan, Edward O. Wilson, Richard Dawkins, and Jacob Bronowski, have made this point quite eloquently. True, science will not promise an afterlife or virgins in Paradise, and it will not explain the cyclic nature of reincarnation or the existence of ghosts. It will also not explain the origin of everything, as it cannot function without a framework of theoretical hypotheses and mathematical constructions. Even if some scientists may not like this, make no mistake: any model that presumably "explains" the origin of the Universe comes chock-full of laws and assumptions built into it, many of them unprovable. As I argued earlier, the very notion that we could possibly construct a theory that explains everything, a final theory, doesn't make sense. Since our tools ultimately define the limits of our knowledge of the natural world, we will never be able to measure all there is to measure. As a consequence, such "final theories" can never be conclusive: there can always be something else out there that has escaped our notice. To accept the limits of science is to accept the limits of human knowledge. Even if we were to create intelligent, self-sustaining machines that would render our existence obsolete, these synthetic mythic beings would still face their own material constraints. They would still need an energy source and would have to face the inexorable march toward disorder as the second law of thermodynamics dictates. In a sense, the ultimate limitation of matter is that, even as conscious mind, it cannot transcend itself.

Science is a way of knowing, of uncovering meaning. It is fed by the same sense of awe that inspires the piety of saints and the deeds of the enlightened, what Einstein called the "cosmic religious feeling." We want to know and we believe that we can. We have faith in our ability to explain the wondrous workings of Nature in ways that make sense to us. When I say that we should abandon any hope of obtaining a final theory, that even a unification of the four forces of Nature based on superstring theories (or their future heir) will never be the final word on the material properties of the world, I am not being defeatist. I am taking the "mind of God" out of science. We don't need a divine purpose to justify our search for knowledge. Science is fine even if forever incomplete. In fact, it's finer for it. Once we accept that science is a human creation and not a fragment of some divine knowledge, we make it stronger, not weaker, as part of our identity as thinking, fallible beings.

Our passion for symmetry has undeniably led to remarkable achievements in the physical sciences. It has also led to big surprises, since Nature has shown, over and over again, that our expectations of perfection are simply projections of our own prejudices. If it weren't for the cold aim of the experimental cannon, our theories would be forever adrift on the seas of perfection. Some of us may not like it, but Nature's mirror is broken: in a perfectly symmetric cosmos, matter would not have coalesced into life forms.

We are masters at building models that explain the workings of the natural world remarkably well. The Standard Model of particle physics is an example of that. So is the Big Bang model of cosmology. Sure, we should strive to obtain better and simpler models, fulfilling the directive of Ockham's razor. Scientists have many fundamental and practical questions to explore. There are also all the questions that haven't been asked yet and that are sure to be. To say that the end of physics or of science is at hand is, in my opinion, sheer nonsense.[6] Equally bad is to believe that if we take from physics the goal of finding a "final theory," all that's left is boring research. I am sure that the vast majority of condensed matter physicists and astrophysicists would take offense at such a statement. Scientists do not need to believe in a "mind of God" behind all of Nature in order to move forward with their search for deeper meaning or to feel awe-inspired.

Nevertheless, the end of a certain way of doing science is needed, a science that places divinelike projections of order and symmetry in a world full of asymmetries and imperfections, a science that searches for ultimate unifying explanations. We could call it the end of the "Ionian Dream." As a disgruntled Eugene Wigner, who won the Nobel Prize in physics for uncovering the symmetry principles behind the quantum mechanics of the atom, once told my senior colleague Robert Naumann, "These Organizing Principles you speak of—they themselves are probably only approximations." We dream up symmetries so that Nature can break them. We should content ourselves with the fact that these symmetries are approximations, that we are modeling what we can measure and what we dream the world to be, not an ultimate reality. Why is this so important? Because it brings science down to the level of the human, freeing it from notions of perfection and ultimate truths.* My point is that there is no Final Truth to be discovered, no grand plan behind creation. Science advances as new theories engulf or displace old ones. The growth is largely incremental, punctuated by unexpected, worldview-shattering discoveries about the workings of Nature. Examples are the dramatic early years of quantum mechanics, a revolution entirely driven by the inability of the established theories of the time to explain new experimental results and, more recently, the discovery of the accelerated expansion of the Universe and the still

* Surely, the truths of physical science remain valid: the force of gravity still decreases with the square of the distance and light still propagates with a constant speed regardless of the observer. Even if the pursuit of scientific questions is related to the cultural context within which it is developed, there is no room for cultural relativism where the laws of motion are concerned. Just try jumping from the top of a building if you don't believe me. (Maybe it's better to just throw a ball down.)

mysterious nature of dark energy, possibly the preamble to a new revolution in physics.

Once we understand that science is the creation of human minds and not the pursuit of some divine plan (even if metaphorically) we shift the focus of our search for knowledge from the metaphysical to the concrete. Given the arresting breadth of what science has achieved in its spectacular explanations of natural phenomena and the challenges that lie ahead, should we not agree that the road ahead will be plenty exciting?

52

MARILYN MONROE'S MOLE AND THE FALLACY OF A COSMOS "JUST RIGHT" FOR LIFE

If you think imperfections are ugly, remember Marilyn Monroe's mole. Would she (and, for that matter, Cindy Crawford and so many other female and male beauties out there sporting facial moles) be more or less attractive without it? Why are these called "beauty marks" if they clearly offend the perfect symmetry of the human face? Facial piercings perform a similar function. You rarely see a person with both sides of the nose or eyebrows pierced.[7]

Several recent studies by cognitive psychologists show that humans prefer slightly asymmetric faces, finding them more attractive.[8] There seems to be a relationship between our aesthetic choices and the left-right functional asymmetry of the brain and of human perception. (The brain is another powerful example of asymmetric functionality, since each hemisphere has very specific and non-overlapping functions.) With the aid of computer technology, we can digitally split the two halves of a face and reconnect them either left side with left side or right side with right side, creating perfectly symmetrical faces. These, it appears, are not as attractive as the naturally asymmetric faces. In fact, as I was writing these lines, I stopped briefly to perform an experiment using the Photo Booth software that comes standard with Apple computers. The result was a very scary left-left Marcelo face, not a photo I would proudly show to family and friends. Readers with access to computers should be able to perform the experiment with their own faces. I am sure that those who have a beauty mark and make it appear on both halves of their faces will not look as good either. The secret of beauty may be in the *slight* breaking of a perfect facial symmetry.

For facial (or body) moles to be considered aids to beauty they must have the right placement and be of just the right size (for example, large dark moles on the middle of the forehead are usually not found very appealing): too small and they are unnoticeable and hence of no use; too large and they are plain ugly. Apparently, the critical size, the boundary between the beautiful and the grotesque, is about one centimeter across. Marilyn Monroe and Cindy Crawford had their moles "just right" and are considered beautiful. There is an aesthetics of the asymmetric here, waiting to be explored some more. But not now. I suggest we take leave of cosmetology, duly inspired, and go back to cosmology.

We have seen how the two standard models of physics, the one dealing with the elementary particles of matter and the one dealing with the Big Bang, rely on a rather large number of parameters (about thirty of them). These include the values for the masses and electric charges of the electron and the quarks, the Higgs mass, the amount of dark matter and dark energy in the Universe, and the amount of matter-antimatter asymmetry. To these we must add the fundamental constants of Nature, which include the speed of light, the constant that sets the value of the gravitational interaction, and Planck's constant h, the constant that sets the magnitude of quantum effects.[9] These fundamental constants of Nature are, in a sense, the alphabet of physics. Every mathematical expression of a physical phenomenon involves, implicitly or explicitly, one or more of them. We could say that the laws of Nature are the grammar rules of the physical sciences, the model parameters and the fundamental constants are the alphabet, and mathematics is the language. For example, the allowed orbits for the electron in a hydrogen atom depend on the mass and charge of the electron, and on Planck's constant h.[10] An extremely important, even if obvious, point is that all constants of Nature have been obtained by careful measurements performed in laboratory experiments. They represent things we don't know how to explain but that are manifest throughout the cosmos. When astronomers examine the spectra of distant stars, they see the same allowed orbits of the electron in hydrogen, in chlorine, and in every other chemical element found here on Earth. This means that the electron has the same charge and mass in stars as on Earth (apart from small and well-understood corrections due to the theory of relativity). Had it been otherwise, had the fundamental constants and model parameters varied from place to place in the Universe, or randomly at different times, it would have been quite difficult to construct meaningful theories of Nature.

The apparent arbitrariness in the values of the fundamental constants of Nature bothers many physicists. Many distinguished colleagues believe that it should be possible for us to derive a more fundamental theory that explains why, for example, the speed of light is 300 million meters per second and not, say, 100 million meters per second. (More precisely, the speed of light is measured to be 299,792,458 meters per second.) This dream theory should, in fact, explain the values of *all* fundamental constants of Nature and *all* model parameters. As we know, this is the goal of unification, to obtain the ultimate explanation for physical reality with no adjustable parameters: the value of every and each one of them should be derived from the theory. According to the best candidate at present, superstring theory, the specific geometric shape of the additional spatial dimensions of space would set the values of the fundamental constants measured in the laboratory. Different geometries would generate different values of the constants.

Briefly, this is how it works. Superstring theories yield better results when formulated in spaces of nine dimensions, as opposed to our familiar three. Since we don't see any evidence of six extra spatial dimensions in our present experiments, they must be so small as to be invisible to us. (Of course, they may also not exist.)[11] The same happens when we see a hose from very far away: it looks like a line (so, a one-dimensional space with only length) although close by it's actually a very long cylinder (so, a two-dimensional space with length and width). The extra six dimensions of superstring models can assume many different shapes, corresponding to different topologies. As a simple illustration, think of a ball with holes punched in it. Balls with different numbers of holes correspond to different topologies: they can't be continuously deformed into one another. In string theory, the extra six-dimensional space can have different topologies, like balls with different numbers of holes. And since, in string theory, the geometry of the extra dimensions is linked to the values of the fundamental constants of Nature, each one of these topologies would correspond to a different kind of world, made of particles with different properties and interactions. According to this view, all physical properties of the particles of matter, their masses and their electric and other "charges," hinge on the specific topology of the extra dimensions of space.

In string theories, even though extra dimensions are invisible to the eye, they essentially define the physical reality in which we live. In the framework of the string landscape, each of its 10^{500} possible geometric realizations would correspond to a different set of fundamental constants, and thus to a universe with very different physics than ours. Perhaps the greatest challenge of present-day superstring models is to come up with a proper selection criterion capable of fetching our Universe from this mess. Or, more ambitiously (and, to my mind, far-fetched), to show that there is only a single theory. An important consequence of these multiverse string models is that the constants of Nature may not have their values set in stone. They could vary from universe to universe, and even in distant parts of our own Universe, although far from what we could observe.

Of course, these ideas are highly speculative and could be completely wrong. We have no clue if superstrings will finally work or if the landscape idea will prevail. Only further research can decide, and even then only if these theories eventually make contact with experiments. Otherwise they will remain "theoretical theories." But as physicists began to ponder the possible variability of the fundamental constants (an idea much older than superstrings), an argument was advanced that I think is misleading, in spite of its widespread popularity. Countless scientific papers and books make the following statement or a close variation thereof: "The values of the fundamental constants define the way the Universe operates [true]. If they were slightly different, say, if the electron charge were 20 percent higher, or if the neutron were not 0.13 percent heavier than a proton but just a little lighter than that, everything would be different. In particular, stars wouldn't shine or even exist, and life would be impossible [true]. If the values of the constants of Nature differed by even small amounts we wouldn't be here [true]. How 'just right' for life is the Universe! [false]." This argument is fallacious in at least two ways. First, it ignores the very important (and obvious) fact that scientists past and present *measured* the constants of Nature through hundreds of experiments and observations. There is absolutely no coincidence here ("just right"), no hidden secret. In *this* Universe at least, they couldn't have had any other values. The existence of fundamental constants of Nature is contingent on the way we construct explanations of natural phenomena, in a progressively more complex and encompassing scientific narrative. It is no wonder that the list of fundamental constants and model parameters has grown in time and, with almost total certainty, will continue to grow. For example, just in the last ten years, we've had to add dark energy and the (still-unknown) values of the neutrino masses. As we probe deeper into Nature and discover new phenomena, we will need new kinds of explanations built upon new fundamental constants.

Second, as I elaborate below, there is absolutely no evidence that our Universe is fit for life. In fact, all evidence at hand justifies the opposite statement, that life exists *in spite of* the harshness and indifference of the cosmos.

When someone says the Universe is "just right" for life, he or she is embracing a very specific agenda: that the Universe is the way it is so that we could be here. This position implies that life and, more crucially, mind (human or otherwise) are cosmically important. Of course we could be tricked into thinking this! Aren't *we* measuring the properties of matter? Aren't we conscious creatures in awe with creation? As we think about the cosmos, the cosmos is reflecting upon itself. Surely this can't be an accident? Perhaps the ultimate reason for us being here is that the cosmos wishes us to be its consciousness?

These are inspiring ideas. But apart from them not having the faintest amount of observational support, I would argue that they are also dangerous.

Let me say it a different way. Someone walks into a library. This person can speak and read only in English. As she inspects the books in the library, she marvels at the fact that they are all in English. "Isn't it amazing that all books here are in English? If they were in different languages, or if some letters were changed here and there, say the letters *a* and *e* into Japanese characters, none of the books would make sense to me. Surely this library was made 'just right' for me to enjoy all these wonderful works of literature! I must really matter." In the strong version of the anthropic principle, this person would conclude that the library must have been built explicitly for her and other English-speaking people.

In the weak version of the anthropic principle, the person would not be surprised by the existence of libraries having books in English. Given that some people can read only in English, some libraries would surely have collections of books in English. There could be books in many other languages in the same library, but only those in English are "just right" for her and her fellow monolinguals. The books in other languages would serve other "typical" readers (that is, literate) that frequent the same library. (So, in this analogy, the library stands for our Universe, with its terrestrial and alien "typical" observers, that is, readers.) There are also libraries with books that don't make any sense at all, say with texts displaying random combinations of letters, or repetitions of the same letter pattern throughout, or simply empty. The person would conclude that such libraries would be devoid of "typical" observers. Within this limited but hopefully suggestive analogy of the anthropic principle, such libraries containing "unreadable" books would correspond to failed universes, that is, to universes unable to harbor typical observers and their meaningful books. The multiverse would resemble the infinite library of Jorge Luis Borges's *Library of Babel*, where all possible books exist and only a few make sense. The strong version is hopelessly self-centered: this book is made for me! The weak version is trivial: in a library frequented by literate patrons, there must exist books in their languages; libraries with meaningless books would have no patrons. Both versions of the anthropic principle teach us much more about the person reading the book than about the library itself.

For a clever fish, water is "just right" for it to swim in. Had it been too cold, it would freeze; too hot, it would boil. Surely the water temperature had to be just right for the fish to exist. "I'm very important. My existence cannot be an accident," the proud fish would conclude. Well, he is not very important. He is just a clever fish. The ocean temperature is not being controlled with the purpose of making it possible for it to exist. Quite the opposite: the fish is fragile. A sudden or gradual temperature swing would kill it, as any trout fisherman knows. We so crave for meaningful connections that we see them even when they are not there.[12]

We are soulful creatures in a harsh cosmos. This, to me, is the essence of the human predicament. The gravest mistake we can make is to think that the cosmos has plans for us, that we are somehow special from a cosmic perspective. We are indeed special, but not because the cosmos has plans for us, or somehow is just right for life. The cosmos couldn't care less for us. Think of the billions, probably trillions, of barren worlds in our galaxy alone. I can't read the message "just right" for life written in so many dead worlds. If the cosmos had plans for us and for cosmic intelligence in general, then the execution of these plans has been very poor indeed. Like the Child God in Olaf Stapledon's brilliant *The Star Maker*, our cosmos seems to stumble ahead, creating and destroying worlds with total indifference: "Thus again and again he fashioned toy cosmos after toy cosmos."[13]

If the constants of Nature are so fit for life, why is life so difficult to find? We are special for being rare, for being alive and conscious of it. The communion we must establish is much more immediate and urgent: it is with our planet and its rapidly dwindling life forms and resources. Our mission is filled with the epic spirit of the chosen, but this is not because we are mythically connected with a purposeful cosmos. Rather, it is because we are dramatically and irreversibly connected with our planet, the only harbor for life that we know.

53 RARE EARTH, RARE LIFE?

In order to strengthen my argument, I want to extend the analysis of Part IV to discuss how rare life and, even more so, intelligent life are in the Universe.

As we study the origin and evolution of life, it is clear that a dramatic increase in chemical complexity and a series of other factors must play out for life to emerge and diversify. Here are some of the most important steps: 1. inorganic chemistry → 2. organic chemistry → 3. biochemistry → 4. first life → 5. prokaryotic cells → 6. eukaryotic cells → 7. multicellular life → 8. complex multicellular life → 9. intelligent life.

Here they are, in more detail:

1. Life needs raw chemical elements, carbon, oxygen, hydrogen, nitrogen, etc. Even if very exotic kinds of life exist out there, the basic chemicals it uses will probably not differ too radically from ours. The good news is that, thanks to supernova explosions, these chemicals are widespread in the cosmos.
2. The raw chemicals must combine into molecules, such as water (H_2O), ammonia (NH_3), and carbon dioxide (CO_2), and then into simple organic molecules, like methane (CH_4) and others. Again, the news is encouraging, as even in the coldness of interstellar space, astronomers have found an extended list of organic molecules, many of them of key importance for life.
3. These organic molecules must find an environment where they can react and become increasingly more complex so as to form biomolecules, the molecules that make up living beings. This is where things start to become more complicated. As we dis-

cussed, water seems to be a crucial ingredient for life to exist. To be sure, we can never discount the possibility that bizarre biochemistries are possible in the absence of water. But lacking proof that they are viable, we must focus our discussion on what we understand and can measure with confidence. The need for a watery environment severely restricts what kinds of celestial bodies can harbor life. The planets must be in the habitable zone of their star, although, as torrid Venus and arid Mars show, this condition alone may not be enough. (Or its definition needs some refining.) Moons, on the other hand, can be in colder regions outside the habitable zone but must sustain enough tidal heating to keep the water liquid, as is the case of Jupiter's Europa. Miller-Urey experiments suggest that the first steps in the chain toward life, the formation of amino acids, may be fairly simple to accomplish as long as the atmosphere and the planet's surface offer the right conditions. However, liquid water and the right chemicals are not enough. For reactions to occur, chemicals must be at high enough concentrations. The planet must be relatively quiescent, that is, not under heavy asteroid bombardment. Its surface must also be somewhat stable, without major tidal deformations and volcanic eruptions.

4. Given all that, the next step is one of the least understood: inanimate chemistry evolved into animated chemistry, a set of self-sustaining chemical reactions capable of absorbing energy from the environment and of replication. Somehow, the one-handedness of life's basic building blocks evolved as well.
5. From this "simple" beginning to the complex proteins and nucleic acids of the first prokaryotic cells, the steps are also blurry. A protective membrane made of fatty molecules surrounded the reacting chemicals, isolating them from the outside environment. With increasing efficiency, the membrane allowed energy and nutrients to come in and waste to get out. Meanwhile, the genetic material inside the primitive cells replicated, leading to fast diversification. This was the world of protozoa.
6. We have little understanding of the next step in life's complexity, the emergence of eukaryotic cells from prokaryotic cells,

although we do know that it took close to 2 billion years. The most accepted view, suggested by biologist Lynn Margulis, is that eukaryotes developed from symbiotic alliances between prokaryotes. For example, mitochondria, the modern cell's little engine, are believed to have been a separate organism in the distant past that was either eaten or absorbed by another cell.[14]

7. Then comes another crucial and equally difficult step, the transition, roughly 3 billion years after life's first known traces, from unicellular to multicellular organisms. As with the transition from prokaryotes to eukaryotes, multicellular organisms possibly also evolved through symbiotic trial-and-error processes, as different kinds of unicellular organisms linked to each other (or ate each other) and became pluralistic in form and function. However, it's hard to understand how the different types of DNA became incorporated into a single genome. As an alternative explanation, the Colonial Theory proposes unicellular creatures grouped in colonies that slowly evolved into multicellular animals. Although the debate is still on, the Colonial Theory continues to gain adherents.
8. Many scientists propose that environmental changes in the Earth played a major role in accelerating the diversity of complex multicellular organisms that climaxed during the so-called Cambrian explosion, about 550 million years ago. Chief among them were the rapid increase in oxygen availability and the advent of plate tectonics and the consequent remixing of surface and ocean chemistry. Tectonics work as a global thermostat, recycling chemicals that help regulate the levels of carbon dioxide and keep the global temperature stable. Without it, surface water would not have remained liquid for billions of years, and life, especially complex life, would have faced insurmountable obstacles.
9. After about 500 million years of evolving multicellular organisms, including many severe mass extinctions and climate changes, the first members of the genus Homo appeared in Africa some four million years ago. Intelligence as we know it is less than 1 million years old. It's been around for less than approximately 0.02 percent of the Earth's history.

Anyone who understands what it takes to climb each of these steps or who has peeked around our own barren solar system cannot possibly state with confidence that life should be ubiquitous in the Universe or, more to the point, that the Universe is "just right" for life. Surely, we *must* search for Earth-like planets, as NASA's Kepler mission is presently doing and the European Darwin mission is planned to do in the near future. Amazingly, astronomers will soon be able to extract information about the chemical composition of these Earth-like planets, searching for telltale signs of life: water, oxygen, ozone, methane, and possibly even chlorophyll. The expectation, which I enthusiastically share, is that life signs will eventually be found. The question is, what kind of life?

Here the divergence starts. In their courageous *Rare Earth: Why Complex Life Is Uncommon in the Universe,* Peter Ward and Donald Brownlee argue very convincingly that life may not be uncommon in the Universe but it likely exists elsewhere only in its simplest form: alien Earth-like planets should support alien microorganisms but not much more than that. Complex, multicellular life depends on too many factors—even after clearing all the chemical roadblocks—to be common. One factor that I haven't yet mentioned is the existence of a large moon. With the exception of Mercury, every planet in the solar system spins about itself at a certain tilt angle, like a wobbling top. Had Earth not had its relatively heavy satellite companion, its tilt of 23.4 degrees with respect to the vertical would have varied chaotically over the years. The consequences for complex life would have been disastrous. A planet's tilt angle determines its seasons and their duration. A variable tilt would render Earth's habitability nearly impossible, as James Kasting from Penn State University stressed in the 1980s. For example, there would be no regular seasons and liquid water would not be a constant presence on Earth's surface for long periods of time. Another important factor is Earth's magnetic field and the protection it offers from lethal radiation from space. Without it, the combination of radiation from outer space and from our Sun would slowly blow the atmosphere away, leaving Earth's surface (and creatures) exposed. This is what happened, for example, with Mars. Although Mars may well have harbored life in the distant past, it's unlikely that it does so now. If it does, it's very well hidden. (Or we don't know how to identify it.) Still, we can be sure only if we keep exploring Mars's surface and subsurface. Perhaps underground protection from nasty cosmic radiation would allow for simple protozoa to survive. Or, as NASA Mars expert Christopher McKay believes, dormant microbes could inhabit the frozen polar regions, where ice is available. This is what happens in Antarctica, where microorganisms have been found in the bottom of frozen lakes in the McMurdo Dry Valleys. To be sure, such life would be a far cry from the little green men of sci-fi, but they would be alien life nonetheless.

The enormous potential payoff of finding any kind of life on Mars would more than justify all the effort being dedicated to the search. The reader may remember the excitement when, in 1996, scientists claimed that a Martian meteorite found in Antarctica, named ALH84001, could be carrying traces of alien life.* The rock had fallen on Earth around 11,000 B.C.E. and had remained under the ice until its discovery in 1984. Longtime NASA administrator Daniel S. Goldin declared: "NASA has made a startling discovery that points to the possibility that a primitive form of microscopic life may have existed on Mars more than three billion years ago."[15] Although Goldin chose his words carefully, a media frenzy followed. Even President Clinton made an announcement at the South Lawn of the White House on August 7, 1996:

> It is well worth contemplating how we reached this moment of discovery. More than 4 billion years ago this piece of rock was formed as a part of the original crust of Mars. After billions of years it broke from the surface and began a 16-million-year journey through space that would end here on Earth. It arrived in a meteor shower 13,000 years ago. And in 1984 an American scientist on an annual U.S. government mission to search for meteors on Antarctica picked it up and took it to be studied. Appropriately, it was the first rock to be picked up that year—rock number 84001.

* How would a Martian rock be found on Earth? The answer is through meteoritic impacts: a violent collision between an asteroid or comet and the surface of a planet kicks enormous amounts of debris into very high altitudes. Some of it may escape into outer space and, after drifting for a while, be captured by a neighboring planet's gravitational field and fall down. Rocks from Earth could also travel to Mars, but Earth's larger mass (and thus gravity) makes it less probable.

Today, rock 84001 speaks to us across all those billions of years and millions of miles. It speaks of the possibility of life. If this discovery is confirmed, it will surely be one of the most stunning insights into our universe that science has ever uncovered. Its implications are as far-reaching and awe-inspiring as can be imagined. Even as it promises answers to some of our oldest questions, it poses still others even more fundamental.

We will continue to listen closely to what it has to say as we continue the search for answers and for knowledge that is as old as humanity itself but essential to our people's future.

Today most (but not all) scientists are convinced that the signs of life found in ALH84001 were not real. One of the methods to search for biotic activity in meteoritic samples is to identify tiny "lifelike" structures embedded in the rocks. The difficulty is that abiotic geological processes can produce signatures that are very similar to bacterial activity. Also, the structures were very small, ten to one hundred times smaller than bacteria found on Earth. Even if one could always argue that Martian life is probably radically different than Earth's, more evidence is clearly needed. Although the case is not completely closed, ALH84001 may not be the proof of alien life we are all hoping for. As far as we know, we remain the only known example of life in the cosmos.

I saw Brownlee at a conference in May 2009 and asked him if he had changed his mind in the past nine years (*Rare Earth* was published in 2000). He has not. If Ward and Brownlee are right, and I believe they are, the consequences are quite serious. Although primitive life forms may not be very rare, Earth-like planets are. And if Earth-like planets are rare, complex life is rare, too. It follows that conscious life, that is, life that is able to reflect upon its own existence, is rarer still, possibly even unique in our galaxy. Instead of stating that the Universe is "just right" for life—implying that life is widespread—we should be amazed that life exists in spite of the harshness of the cosmic environment. Could we then be alone in the cosmos? Or is the Universe filled with intelligence? We must now turn to the possibility of intelligent life in the cosmos and examine it under the light of what we have learned.

54 | US AND THEM

This title is borrowed from a comparative literature course I occasionally teach at Dartmouth, where I examine how the thinking about aliens and, in particular, about extraterrestrial intelligence, has changed in Western culture since the seventeenth century.[16] "They" are a projection of our fears and hopes, a mirror image of humanity's best and worst. In most stories and movies, aliens' looks and machines directly correlate with the state of the science and technology of the time. At the close of the nineteenth century, H. G. Wells's *War of the Worlds* pictured Martians coming to Earth via powerful detonations resembling cannonballs or missiles. Soon after, with the progress of human flight, aliens began to fly as well. With the development of genetics and nuclear physics, mutations and nuclear power started to appear; the same happened with computers from the 1950s onward. In many stories, aliens are able to do what we only dream of doing. In Arthur C. Clarke's masterpiece, *2001: A Space Odyssey*, the aliens are indistinguishable from gods, "creatures of radiation, free at last from the tyranny of matter."[17] Imagine what a person from Copernicus's time would think if shown a laptop computer or an iPhone. Even my father, born in 1927, would have been incredulous. He was already suspicious of VCRs in the 1980s.

The possibility that alien life exists is tantalizing. Finding even a single extraterrestrial microbe would be perhaps the greatest of all scientific discoveries. The world would never be the same. If we had conclusive evidence that life emerged independently somewhere else, the supposition that it should be widespread in the cosmos would be greatly strengthened. Life would not be a terrestrial anomaly. Given that the laws of physics apply across the Universe and that the same chemical elements are found in other stellar systems, it follows that if we found primitive life in at least one other planet or moon in our cosmic neighborhood, we should indeed expect it to be widespread: the principle of mediocrity, at least as it relates to the *existence* of life, would gain support. The cosmos could indeed be biofriendly, and the possibility that there is a deep link between life and the cosmos would have to be taken seriously. As Paul Davies eloquently wrote, "If life follows from soup with causal dependability, the laws of nature encode a hidden subtext, a cosmic imperative, which tells them: 'Make life!' . . . This is a breathtaking vision of nature, magnificent and upliting in its majestic sweep. I hope it is correct. It would be wonderful if it were correct."[18] A longing for cosmic companionship joins the longing for the harmonies.

Proof of a "cosmic imperative" for life would be a blow to the Darwinian orthodoxy that discounts any kind of determinism in life-related processes. According to the theory of evolution, there is no purpose or plan for life; the drama of existence unfolds as creatures struggle to survive in challenging environments. Hence it would be very hard to resist the temptation to equate some kind of "cosmic imperative" for life with a sense of purpose, with the existence of a divine universal quality that favors (or even creates) life. On the other hand, even though the discovery of life elsewhere would be deeply transforming for humanity's sense of identity, there is no need to equate it with cryptoreligious notions. For example, one could equally argue that there is a cosmic imperative for stars, given that stars are all over the place. Yet we know that stars result from the gravitational contraction of clouds of hydrogen gas, a (fairly) well-understood physical process that, under the right conditions, is repeatable across space without any purposeful intentionality. In other words, even if primitive life were discovered elsewhere and we were to conclude that it was fairly common in the cosmos, we could keep the explanation for its existence well within scientific boundaries.

The situation would be quite different if alien life were multicellular. It is one thing to find alien amoebas; it is quite another to find complex alien life forms, with distinct organs serving specialized metabolic and motor functions. Given the remarkable resiliency of terrestrial extremophiles (see Part IV), I am quite confident that microbial life should not be very rare. However, I am doubtful that complex multicellular life should be as common. Quite the opposite. Given the enormous obstacles terrestrial life had to overcome in its path toward multicellular complexity, and given the overall barrenness of our own solar system (in spite of its glorious beauty), how common should we expect complex life to be? Of course we don't know the answer, and all we can do for now is speculate. But from what we can tell so far, the odds are clearly not very good.

The more complex life becomes, the more fragile it is. Complex animals cannot withstand large temperature fluctuations or live in extreme temperatures or environments, as some bacteria do. Their host planet must have a very stable thermostat, a condition that implies a series of geological and atmospheric constraints, as we saw above. Also, the larger the creature the more vulnerable to attacks, the more energy it needs to survive, and the harder it is for it to adapt to sudden environmental changes. Since it's difficult to imagine how intelligence—here or anywhere else—could have emerged without millions of years of evolving multicellular creatures, the discovery of multicellular aliens would be a great boost to the belief that there are other smart creatures out there. Even so, it's important to keep in mind that human intelligence appeared as a by-product of random cosmic and evolutionary accidents: intelligence is *not* the end goal of evolution, as 150 million years of dinosaurs demonstrates. But I would be the first to agree that the discovery of complex alien life would indeed be revolutionary.

In 1960, radio astronomer Frank Drake came up with a quantitative way to evaluate the probability of our galaxy harboring intelligent life. His strategy, known as the Drake equation, was to essentially multiply the various factors needed in order for intelligent life to exist in a stellar system. The advantage of the equation and its modern versions (as for example, discussed in *Rare Earth*) is that it helps us understand what it takes for intelligence to emerge on a planet. The disadvantage is that we don't know how to estimate most of the terms in the equation with reasonable precision and even what terms to include. For example, although we know the number of stars in the Milky Way fairly well (between 200 billion and 400 billion), we don't know what fraction of them have planets in the habitable zone, how many of these harbor life, how many harbor complex life and, finally, how many harbor intelligent life. Also, should we include in the equation terms quantifying the fraction of planets with a large moon, with plate tectonics, and with radiation-shielding magnetic fields? The choice of what terms to include in the Drake equation and how to estimate them has a lot to say about the intentions of who is including them. Carl Sagan, for one, offered an early estimate that maybe 1 million civilizations in the galaxy are capable of radio communications. Others claim the number is one: us.

If we take the possibility of alien intelligent life seriously, we must consider a series of consequences. The most obvious, as Enrico Fermi had already considered in 1950, is "Where is everybody?" Our galaxy is about 13 billion years old, more than twice the age of the Sun. If we imagine that life evolved in another stellar system, and that it reached a stage in its evolution where complex creatures became intelligent even as little as a few million years earlier than it did here, then it follows that some of these aliens would have had plenty of time to reach amazingly advanced levels of technological sophistication. Considering what we have achieved with only four hundred years of modern science, their technology would be like magic to us. If, like humans, they suffer from wanderlust (what do we know of aliens' psyches?), they would have had the means and plenty of time to explore the galaxy many times over. Yet at least from what we can tell, they haven't colonized the galaxy or visited us here on Earth. So, where is everybody? This issue is sometimes called Fermi's Paradox.[19]

One answer is that they came here and left without a trace. (Not very useful as proof of alien visitation.) Another is that they came here long ago and implanted life themselves: Earth is a zoo for the aliens, a laboratory for evolutionary biology. To us, as to the earthlings in Kubrick's *2001: A Space Odyssey,* these aliens would be indistinguishable from gods. Another answer to Fermi's Paradox is that our lives are a virtual animation. As in the movie *The Matrix,* we are all prisoners, victims of a fantastically complex simulated illusion we call life. Another is that they are here, but we can't see them due to their cloaking devices. (Also not very useful as proof of alien visitation.) A more serious argument, clearly a product of the Cold War, is that no technological civilization is capable of surviving past the nuclear age. As explored in the classic sci-fi movie *The Day the Earth Stood Still* (1951), young intelligences like ours are too morally immature to handle such power. Likewise, most (all?) alien intelligences would be too immature to handle the power of self-destruction. We don't see them because they don't exist any longer. Although appealing, this argument is too dramatic: it's hard to envision a civilization reaching such destructive power as to completely obliterate life from a planet. (Of course, we cannot discard the possibility that, as in *Star Trek,* some civilization could have found a way of imploding a whole planet with "red matter" or the likes of it. But I will stick to what we think is reasonable for now.) Even after such horrendous cataclysm, some creatures would survive, and odds are that intelligent ones would be among them. Any form of life comes coded with some version of the survival instinct. If they are anything like us, they would start again and rebuild, hopefully on a more peaceful path. Unless, of course, the devastation reaches the hopeless levels described in Cormac McCarthy's *The Road,* where, apart from a few humans, Earth is literally dead.

The fact is, even if "they" are out there, we may never know. For practical purposes, until we learn otherwise—and it could take a very, very long time—we are alone. And this realization should change the way we think about ourselves and about the world we live in.

55 | COSMIC LONELINESS

In spite of the excitement that intelligent aliens and UFOs arouse, current science tells us that we are it, probably for the long run. Unless radio astronomers working at the Search for Extraterrestrial Intelligence (SETI) find, against serious odds, uncontroversial evidence of intelligent life elsewhere, we must face the fact that we are—for all practical purposes—alone. We have explored some of the arguments pointing to the rarity of life and, even more, to that of complex intelligent life. Apart from all the geophysical and biological obstacles, even if such creatures exist, they are lost from us in the remoteness of space. As an example of how hard it is to travel across interstellar distances, a trip to Alpha Centauri, the nearest star to the Sun, would take about one hundred thousand years with our fastest spaceships. And that's only to our nearest neighbor! Even if we find ways to travel at, say, one-tenth of the speed of light, such a trip would take forty-five years to complete. Sure, we could imagine constructing self-sustaining biospheres populated with enough humans and terrestrial flora and fauna to perpetuate life for thousands of generations until they reach other solar systems. But from the perspective of a short human life, physical contact with aliens is highly improbable.

I agree wholeheartedly with SETI enthusiasts that we have a chance of finding intelligent alien life only if we look for it. Hints could come from radio signals, from alien space debris, and even from gigantic astronomical engineering projects. (With apologies to those who believe aliens are already here or come often, a visitation from "them" is nearly impossible to contemplate.) As with the search for any sign of life in our own solar system, the payoff for finding evidence of intelligent alien life is so tremendous that it's worth pursuing. It would indeed be wonderful if SETI succeeded; the world would never be the same again, as Carl Sagan's novel *Contact* so brilliantly conveyed. We would know that we are not alone in the Universe, that other minds are out there, contemplating the mysteries of existence. Still, communication would be a hopeless task. Due to the limiting boundary of light's speed, we wouldn't be able to maintain a conversation. Imagine that smart aliens inhabit a planet orbiting Alpha Centauri. Even if we accept that a reasonable communication channel or language could be established (as math was in *Contact* and music was in Steven Spielberg's movie *Close Encounters of the Third Kind*), conversation would not be very lively. The time between each message and its response would be nine years. Optimistically, the initial message would be sufficiently long and complex to keep us busy until the next one arrived. But in all honesty, after fifty years of SETI and ten thousand years of civilization, what we have is complete cosmic silence and a total absence of convincing evidence of alien visitation, a solar system devoid of any obvious form of active life, and stellar systems that are remote and probably very unlike our own. Even if intelligent life forms do exist somewhere in our galaxy, we may never hear from them or know of their existence. Even if we are not alone, it sure feels like it.

56 A NEW DIRECTIVE FOR HUMANITY

After five millennia of intense searching and hoping for some kind of ultimate explanation for everything, religious or scientific, we must move on. True, the search has taken us into new realms of knowledge, as we uncovered some of Nature's deepest secrets. We have given wings to our imagination, creating magnificent works of music, literature, and art to express our longing for understanding and for each other. The very first human died entranced by the mystery of existence. The very last will as well. We will keep on searching and creating. But our focus must change. Science has shown us that reason, fueled by a passion for discovery, is our most powerful tool to answer questions about the natural world. Given how our first musings about origins emerged in mythic imagery, it should not be surprising that science carries, in its roots, the same mythic longings for ultimate explanations. Yet Nature is telling us otherwise.

In spite of our longing for the harmonies, Nature tells us that its creative power comes from asymmetries writ deep into the world, from the very small all the way to the very large. We hope for perfect symmetry, write powerful equations to express it, and find that our solutions are only approximations to an imperfect reality. That's how it should be. From asymmetry comes imbalance, from imbalance comes change, from change comes becoming, the emergence of structure. Some of the fundamental symmetries of particle physics must be violated in order for matter to exist. The Universe as a whole may have emerged from a quantum fluctuation out of some timeless realm where many universes coexist: a fluke that carried the seed of existence. Out of randomness, the cosmos evolved to generate the lightest chemical elements. Clouds of hydrogen, cloaked in invisible dark matter, collapsed under their own self-gravity to form the first stars and galaxies. Billions of years later, around one star, a watery planet collected the ingredients that would make life possible. After much turbulent mixing and cataclysmic collisions, the planet quieted and, out of the primordial ooze, molecules grew and bonded to coalesce into the first living creature. Billions of years afterward, our ancestors began to contemplate Creation. Alone, they looked at the skies with fear and awe.

We have learned much about where we are and what we are made of. Our amazing tools have greatly extended our vision of the world and of the Universe beyond. We have also learned that there is much that we don't know, and much that we will never know. Our reach is wide but limited. Nature need not comply with our flights of fancy. Science tells us how Nature works and not how it should work. As we left our home planet, sending probes across the solar system, we were struck by how different these other worlds were, how magnificent and barren, how indifferent to life. We have longed for the harmonies for too long; we have longed for cosmic companionship—divine or alien—for too long. We must accept that we are alone in the cosmos, if not in absolute terms—as we can never be certain of what lies beyond our instruments—at least in practice. This makes us very special indeed. And creates a new purpose for humankind.

Some may accuse me of trying to restore some kind of anthropocentric view. (*Humancentric* would better serve my purpose.) They are right, although certainly not if they think that my anthropocentrism relates to that of pre-Copernican times. I *am* saying that we are unique and important. But not for having been created by a god, or for being the result of a purposeful cosmic directive. We are unique and important for being alive and self-aware. All that we know and will probably know for a long time indicates that we are the only ones asking questions. We may not be the measure of all things but we are the only things that can measure. The acceptance of our cosmic loneliness is a wake-up call, ringing to arouse a new consciousness. Humans! Wake up and save life with all that you have! Life is rare. Treasure it, worship it, make it last, spread it across the Universe. This is our supreme mission as the minds of the cosmos.

The timing for this revelation couldn't be more urgent. The fast pace of progress, the promise of riches and of a better life, has left us blind to the damage we inflict on our planet. Yes, we must survive, plant, build, and explore Earth's resources. But we cannot continue at our current pace, indifferent to the devastation we are inflicting on our planet and on the precious life it harbors.[20]

The climate is changing and species are dying at the alarming rate of roughly thirty thousand per year. We are witnessing the greatest mass extinction since the demise of the dinosaurs 65 million years ago. The difference is that for the first time in history, humans, and not physical causes, are the perpetrators. We destroy habitats, pollute rivers, hack down mountains and forests, flood valleys, introduce species to new environments without planning, kill, fish, and hunt endangered species with impunity. In our frenzy, we forget that Earth is limited in its resources and its resilience. Life recovered from the previous five mass extinctions because the physical causes eventually ceased to act. Unless we understand what is happening and start acting together as a species we may end up carving the path toward our own destruction. We will only effect positive change on a planetary scale when we truly comprehend the value and precariousness of our situation in this pale blue dot. Unfortunately, extinctions and climate changes usually happen on time scales much too long to be seen in a typical human life span. We don't "see" threatening effects happening fast enough to be scared. There is no gun to the head prompting us to react. How close must the collapse of our planet be before we are scared enough to mend our ways? How long are we willing to wait until we are convinced that change is needed?

I know I sound like a doomsday prophet, and I don't like it. I wrote a whole book trying to clarify the connection between science and apocalyptic prophecies.[21] Some may shrug these lines away, and go on with their business. But I hope not too many. I hope that once they realize how rare Earth is, how rare complex life is, how precarious and how precious our existence is, they will embrace the cause for survival. We need a new morality aimed at preserving life here and, one day perhaps, at spreading it across the cosmos. But for each one of us, the work starts in our backyard.

The most amazing fact about existence is that we are aware of it. The most sobering is that, as with our ancestors, we remain alone as we contemplate Creation. Since, tragically, only wars and common foes unite nations, let us come together as a species to fight for life. But unlike our past and present wars, this is not a war of boundaries or creeds. This war is between our past and our future. And it can only be fought in the present.

EPILOGUE

GARDEN OF DELIGHTS

When I was a little boy, I used to live in an enchanted garden. I wasn't there always, as the garden was in my grandparents' house in the mountains outside Rio de Janeiro. But when summer arrived, and I knew it was time once again to pack our bags and go to Terezópolis for three months, everything got easier. In the big old house, the darkness was my friend, not my enemy. There were no monsters lurking in the shadows, no fangs about to pierce my neck as there were at home. At night I would go outside with my cousins and lie down on the lawn to see who could count more shooting stars. When it was cloudy, I would find a long bamboo stick and go hunting for bats, waving the stick in the air. The poor creatures would try to track the stick's motion and end up colliding with it. I wanted to see them close by, these friends of vampires (and some, vampire bats themselves).

Once, on a good night, I managed to catch two bats and one survived. Boys are cruel. But at eight or nine, I was busy exploring a world alien to me, a world that was alive with all sorts of creatures I never saw in my apartment in Rio. Going to the old house was like entering a parallel universe, where a new Marcelo, brave and curious, lived. I put both bats in a large can and went on with my business, trying my luck with frogs. After a while, I decided to check on the bats. The one that had survived was on top of the dead one, fangs deep into the back of its neck. The vampire bat was trying to survive the only way it knew how. I was fascinated. In my excitement, I went running to my grandmother to show her my catch. The old Ukrainian lady almost collapsed. What was to me a clear demonstration of the fierceness of Nature (and of my bravery) was to her a disgusting sight. I took the bats outside, buried the dead one, and released the other into the woods. From then on, I would keep my bounty to myself.

The days were as magical as the nights. Growing up in the tropics is a blissful portal into the natural world. Life exploded in all forms and sizes, from the tiniest spiders to huge iridescent-blue butterflies, from countless kinds of orchids, hyacinths, hibiscus, and giant ferns to birds of all colors. With magnifying glass in hand, I would saunter about the gardens, stopping to examine anything that moved. Life was definitely squishy here! But I had more sense than to just squash my bugs. I collected them. Dozens of glass jars filled with specimens of my bug collection preserved in alcohol lined the walls of my bedroom: spiders, ants, beetles, bees, wasps, millipedes, amazing praying mantises, the rare walking stick, and so on. As soon as I found a new species, I would try to locate it in my books and classify it, carefully labeling the jar. I wanted Nature to be part of my life, not just an accessory to hold at a distance and admire once in a while. There was no place I was happier than among these gardens. My eyes filled with tears every time I had to return to the big city. (In spite of the well-deserved reputation of its fabulous beaches and beauty, Rio is a metropolis of some 10 million people.)

The day the old house was sold, something died inside of me. How could I accept that, from then on, that parallel universe of magical life was to live in my memory only? After a few years, my parents bought another house in Terezópolis. And although it was a nice house, things were never the same again. To make things worse, to get to the new house we had to drive past the old one. And every time we drove by the old house, I noticed that a little more had been taken away from her. The garden was the first to go; then the magnolia trees; then the grass fields. I was told that the house had become a seminary. Apparently the Bible students had no time or inclination to care for earthly things. As in Poe's *The Fall of the House of Usher*, the old house had died, too.

I tell this story because it relays a deep sense of paradise lost, a loss not dissimilar to what we may experience in the coming decades. We have a chance to change the course of things and salvage the world we grew up loving. Even if some have doubts as to how severe the upcoming storm will be, there will be a storm. The first raindrops are falling already.

We should not be gambling with our children's future. I have four of them and hope to see their children sauntering in my gardens one day, magnifying glass in hand, in blissful awe at the wondrous squishiness of life.

NOTES

Part I: Oneness

1. In a way it does that, when electrical discharges jolt stopped hearts back into pumping mode. Yet the question as to why some hearts get revived while others refuse to beat again remains unanswered. If doctors knew, they wouldn't invest so much emotional and physical energy trying to revive the ones that won't. The boundary between life and death remains obscure.
2. http://religions.pewforum.org.
3. http://www.adherents.com/largecom/com_atheist.html.
4. http://www.astarte.com.au/html/pella_s_canaanite_temple.html; http://cogweb.ucla.edu/Culture/Monotheism.html.
5. Gerald Holton, "Einstein and the Goal of Science," in *Einstein, History, and Other Passions* (Cambridge, Mass.: Harvard University Press, 1996), p. 161.
6. Isaiah Berlin, "Logical Translation," in *Concepts and Categories* (New York: Viking, 1979).
7. Readers interested in exploring in more detail the origins and perpetuation of the Pythagorean myth are referred to Charles Kahn's *Pythagoras and the Pythagoreans: A Brief History* (Indianapolis, Indiana: Hackett, 2001).
8. An enthralling history of the book and its readers across the centuries can be found in Owen Gingerich's *The Book Nobody Read* (New York: Walker, 2004).
9. It is possible that Maestlin's true intention was to protect Kepler from his more radical colleagues. We will never know.
10. The details are quite complicated, but as a rule, the fitting worked to within an accuracy of 5 percent or better. Interested readers can consult E. J. Aiton's introduction in the 1981 translation of the *Mysterium* published by Abaris Books, New York.
11. Holton, *Einstein, History, and Other Passions*, p. 160.

Part II: The Asymmetry of Time

1. Many books, including my own *The Dancing Universe*, tell the history of the Big Bang model in detail. In the bibliography I list a few.
2. Penzias and Wilson got the Nobel Prize in 1978 for the detection of the cosmic microwave background in 1965. Sadly, Gamow died in 1968, hardly with

enough time to celebrate. I met Alpher in 2005 in his retirement home in Tampa, Florida. Even his gentle demeanor couldn't hide the enormous disappointment of having his work practically forgotten. On July 27, 2007, Alpher's son received the National Medal of Science from President George W. Bush on his behalf, the highest scientific honor in the United States. Alpher passed away sixteen days later.

3. Tragically, people still kill and die for their gods, as they have for millennia. Their motto continues to be "In (our) God we trust."
4. In *The Prophet and the Astronomer: A Scientific Journey to the End of Time*, I explain the several stages of a star's life in detail. I also list other references in the bibliography.
5. Neptunium, element 93 (93 protons in its nucleus, its *atomic number*), and plutonium, element 94, are found in uranium ores, albeit in exceedingly small traces. Elements from americium, atomic number 95, and higher are made artificially in laboratories.
6. Even though, strictly speaking, a proton is not a ball with a definite radius, we can associate with a proton a length scale called its Compton wavelength, which is about 0.13 trillionths of a centimeter, or 1.3×10^{-13} cm.
7. Readers interested in learning more about the string landscape should consult the book by Leonard Susskind, *The Cosmic Landscape: String Theory and the Illusion of Intelligent Design* (New York: Little, Brown, 2006).
8. A description of the competing theories of quantum gravity can be found in Lee Smolin's *Three Roads to Quantum Gravity* (New York: Basic Books, 2001).
9. Why not just 14 billion light-years? This would have been the answer if space were not expanding. But because it is, the photon gets a boost from the space stretching behind it like a surfer being carried along by a wave: in the same amount of time, the surfer (and the photons) can travel farther. For the photons, we can see all the way to three times as far.
10. Exponential growth is very fast. Imagine you have a one-yard ruler. If it were to be stretched exponentially fast in time, in 1 second it would be 2.72 times longer; in 10 seconds, it would be 22,026 times longer; in 60 seconds, it would be about 100 trillion trillion times longer (to be precise, 1.14×10^{26} times longer). Guth's excellent book *The Inflationary Universe: The Quest for a New Theory of Cosmic Origins* (Reading, Mass.: Addison-Wesley, 1997) tells the story of inflation in detail. A more recent and also very readable reference is Alex Vilenkin's *Many Worlds in One: The Search for Other Universes* (New York: Hill & Wang, 2008).
11. Physical theories are easier to analyze under this light; we know that Newtonian mechanics fails for speeds close to the speed of light and for atomic distances. In contrast, it is difficult to imagine when Darwin's theory of evolution might fail.
12. I find the term *negative pressure*, even though mathematically correct, to convey a very confusing image. After all, positive pressure makes balloons

expand. Negative pressure would seem to make something collapse, exactly the opposite of what happens in cosmology. Recall that the smaller the pressure the faster the expansion. Back to our pressure-as-"mass" analogy, the trick in inflation is to find a kind of matter that generates negative pressure, as if it had "negative mass": contrary to normal matter with zero or positive pressure and that tends to collapse onto itself due to its own gravity, the stuff driving inflation forces space to expand ultrafast, causing it to stretch out as efficiently as possible. Some authors like to call it antigravity, although the term is misleading. Gravity is still perfectly attractive. The repulsive effect acts only on the geometry of space.

13. As I was writing these lines, the 2008 Nobel Prize in physics was announced. The three winners—Yoichiro Nambu, Makoto Kobayashi, and Toshihide Maskawa—pioneered precisely the notion that particles of matter may have different properties and behavior at different temperatures and energies. Nambu's inspiration, in particular, was precisely the qualitative changes that occur with ordinary substances such as water and metallic alloys as temperatures are raised and lowered past a critical value.
14. There are many names connected with inflation, even in its early days. At the risk of offending a few colleagues for my omission (and I apologize beforehand), here are a few of the key players during the early 1980s: Andrei Linde, now at Stanford; Andreas Albrecht, now at the University of California at Davis; Paul Steinhardt, now at Princeton University; Alexei Starobinsky, at the Landau Institute in Moscow; Stephen Hawking, from Cambridge University. See the bibliography for references on cosmology and inflation.
15. In 2006 I tried rescuing the original inflationary scenario, but my model is also quite tentative, using two scalar fields instead of only one. Other more notable attempts by many of my colleagues suffer from similar problems. Inflation is an idea in search of a compelling model.
16. Moons can also receive a slight illumination from reflected starlight coming from their planets, a phenomenon known as planetshine. This is why we sometimes "see" the dark disk of the crescent Moon. Leonardo da Vinci was the first to propose this explanation.
17. An example: http://www.spacetelescope.org/images/html/heic9910b.html.
18. In Geoff Brumiel, "A Constant Problem," *Nature* 448 (2007): 245–48.

PART III: THE ASYMMETRY OF MATTER

1. Steven Weinberg, *Dreams of a Final Theory* (New York: Pantheon, 1998), p. 148. The same idea was recently expressed by Frank Wilczek in his *The Lightness of Being* (New York: Basic Books, 2008), p. 136. From the screenplay of *Amadeus*, Wilczek quoted Salieri's comment on Mozart's music: "Displace one note and there would be diminishment. Displace one phrase and the structure would fall."

2. This is strictly true only for the highly idealized case that doesn't include complications from external fields, relativistic corrections to the motion of the electron about the nucleus or about itself (its spin), and many others. Ingenious and highly successful approximation methods have been invented to study the effects of these complications.
3. The reader may be wondering why wouldn't all matter and antimatter annihilate if the symmetry were exact. The reason can be traced back to the expansion of the Universe. As the Universe expands and grows, not all particles and antiparticles can find each other to annihilate. A population of wanderers is left over. This cosmological process is called *freeze-out*. It determines the relative abundances of relics from early cosmic history.
4. I list several titles in the bibliography.
5. M. Gell-Mann, "A Schematic Model of Baryons and Mesons," *Physics Letters* 8 (1964): 214.
6. David Lindley, *The End of Physics: The Myth of a Unified Theory* (New York: Basic Books, 1993).
7. Although Juan Sebastián Elcano, Ferdinand Magellan's second in command, completed the first circumnavigation of the globe only in 1522, the sphericity of the Earth was common knowledge throughout the late Middle Ages and the Renaissance. In fact, it has been common knowledge since Ancient Greece and was proven around 240 B.C.E. by Eratosthenes, the third librarian of the famed library of Alexandria.
8. We addressed magnetic monopoles in Part II.
9. Pauli actually called the particles neutrons. When Chadwick discovered *the* neutron in 1932, Fermi suggested, in wonderful Italian fashion, the name *neutrino*, the little neutron.
10. For the more technically inclined: there is a possible small violation of lepton number conservation in the Standard Model via the so-called *chiral anomaly*, a purely quantum effect.
11. Unless, that is, you are one of those rare people with *situs inversus*. In that case, your heart and all your major organs are mirror-reversed: heart and stomach on the right, liver and appendix on the left. Their mirror image has the heart on the left, just like most of us.
12. More precisely, a particle's spin is given in multiples of $\hbar/2$, where $\hbar$ is the so-called Planck's constant h divided by 2π. Planck's constant, which is a tiny number, sets the scale for quantum mechanical effects. Quantum systems have only a few allowed states, such as the two spin states of quarks and leptons: a jump between states is a dramatic event. Classical systems, like a top, have such a huge number of allowed states that their behavior seems continuous. Think of a ball rolling down steps (a bumpy discontinuous motion) or down a ramp (discontinuous motion).
13. Baryon number (B) is more appropriately defined in terms of the number of quarks in a hadron: B = (number of quarks – number of antiquarks)/3. So, mesons have zero baryon number (one quark and one antiquark), while

protons and neutrons have baryon number = +1 (three quarks and zero antiquarks).

14. Dan Hooper, in *Nature's Blueprint* (New York: HarperCollins, 2008), introduces supersymmetry in detail and explains why it's so attractive to so many.
15. As of this writing, in the fall of 2009, there was no compelling way to incorporate neutrino masses within the current formulation of the Standard Model. Many view this fact as an indication that the model is incomplete, which it surely is. How to complete it, or whether it can be completed, is still an open question.
16. Vadim A. Kuzmin, Valery A. Rubakov, and Mikhail E. Shaposhnikov, "On anomalous electroweak baryon-number non-conservation in the early universe," *Physics Letters* 155B (1985): 36.
17. For the more technically inclined, the electroweak theory has two coupling constants: one for the symmetry (group) related to the weak force (g) and another for the symmetry (group) related to electromagnetism (g'). For example, what we call the electron charge is a mixture of these two, $e = g.g'/(g^2 + g'^2)^{1/2}$. Against the spirit of unification, there are always two symmetry groups (and their related coupling constants) and not a single one that encompasses the two interactions of the low-energy theory.

PART IV: THE ASYMMETRY OF LIFE

1. Still, these experiments pale when compared with the public electrocution of Topsy the elephant, masterminded by Thomas Edison. The female—which had killed three humans, including an abusive keeper—was shocked with 6,600 volts fed from an alternating current and died in seconds. Edison staged the event to scare people away from alternating currents, the rival to his direct current. He called it being "Westinghoused," referring to George Westinghouse, who, with Nicola Tesla, had invented alternating currents.
2. An early account of the curative power of electricity can be found in Reverend John Wesley's remarkable *The Desideratum, or Electricity Made Plain and Useful*, published in 1759. The reverend minced no words when describing his faith in the new "cure": "I doubt not, but more nervous disorders would be cured in one year, by this single remedy, than the whole *English Materia Medica* will cure, by the end of the century." See books.google.com/books?id=Wx4DAAAAQAAJ&pg=PA9&dq=electricity+and+the+soul.
3. From the introduction to the third edition of Mary Shelley's *Frankenstein: Or, the Modern Prometheus*, published in October 1831.
4. See, for example, www.snopes.com/religion/soulweight.asp and Mary Roach's *Stiff: The Curious Lives of Human Cadavers* (New York: Norton, 2003). Dr. MacDougall's measurements inspired the 2003 Hollywood hit movie *21 Grams*, which featured Sean Penn playing the role of an ailing mathematician.

5. A thorough account of the Miller-Urey experiment and its impact on astrobiology can be found in Christopher Willis and Jeffrey Bada, *The Spark of Life* (New York: Perseus, 2000).
6. This theory of Moon's origins has plenty of supporting evidence: the composition of oxygen isotopes on Earth and Moon are very similar, indicating that the Moon formed near the Earth; also, the Moon lacks iron, which is plentiful on Earth, especially near its core. This can be explained if the impacting body hit asymmetrically, dislodging mostly material closer to the surface, leaving Earth's iron-rich core intact. Other theories fail to explain these findings.
7. Many researchers, most notably David Deamer from University of California at Santa Cruz and Jack Szostak from Harvard University, have done remarkable work studying the properties of the simplest possible cells. While Deamer investigates how lipid (fats) boundaries form around genetic material, Szostak tries to zero in on the simplest possible cell—with the smallest amount of genetic information—still capable of being deemed as living.
8. The Moon is still moving away from us, albeit at the much humbler rate of about three to four centimeters a year. Consequently, tides are weakening.
9. More technically, the synthesis of deoxyribose—the carbohydrate backbone of DNA—takes place via ribose—the backbone of RNA—indicating that RNA is in a sense more primitive than DNA.
10. Tom Fenchel, *Origin and Early Evolution of Life* (Oxford: Oxford University Press, 2002), p. 51.
11. Louis Pasteur, *Researches on the Molecular Asymmetry of Natural Organic Products* (Edinburgh: Alembic Club, 1905), p. 10.
12. Ibid., p. 19. Pasteur used the technical word *hemihedral* instead of asymmetric. Hemihedral crystals are those with half the number of planes required for full symmetry.
13. There are many important practical consequences of chiral asymmetry, as the pharmaceutical companies know well. Left- and right-handed compounds can have very different medicinal effects. A tragic example is that of thalidomide. One form is an effective medication against morning sickness in pregnant women; the other is teratogenic, causing horrible birth defects. In the 1950s and '60s, many children around the world were born severely deformed because their mothers were given the wrong form of the medicine. Another example is ethambutol: one forms treats tuberculosis while the other causes blindness. On a lighter note, in Lewis Carroll's *Through the Looking Glass*, published in 1871, Alice is well aware that foods with the wrong chirality can have bad effects: "How would you like to live in the Looking-glass House, Kitty? I wonder if they'd give you milk in there? Perhaps Looking-glass milk isn't good to drink."
14. Pasteur, *Researches*, p. 42.
15. Ibid., p. 40.
16. Plasson and collaborators use a (nonautocatalytic) mix of left- and right-handed chiral compounds with a small initial chiral bias. As in Frank's reac-

tion, this small bias is successfully amplified, resulting in a solution of small chains of chirally pure compounds called peptides. In this model, each little white or black pearl has a tiny lightbulb attached to it that may be either on or off. If it is on, it means that the pearl (the amino acid) is active and ready to link up with another pearl. The activation of the amino acid (the turning on of the lightbulb) depends on external energy sources, typically compounds such as nitrogen oxide or carbon monoxide that are the chemical equivalents of granola bars (energy-giving foods) to people.

17. Very roughly, in a volume containing N molecules, we should expect a fluctuating excess of $\sqrt{N}$ molecules of one type or the other. For example, if we have 10^{24} molecules, we should expect an excess of roughly 10^{12} either left- or right-handed molecules, or 1 trillion. This may seem like a lot but it really isn't. If N were the world population, 7 billion, $\sqrt{N}$ would be only about 84,000 people.
18. Pasteur, *Researches*, p. 43.
19. This crucial correlation is still poorly understood. There have been a few experiments by Sandra Pizzarello from Arizona State University and Arthur Weber from the SETI Institute that indicate that left-handed amino acids can catalyze the production of right-handed sugars. At the time of writing, I am trying to model this effect with Sara Walker. The results are very promising. Left-handed amino acids—either made on early Earth or rained down from the sky—could have jump-started the polymerization reactions, biasing the formation of right-handed sugars. In this case, the stage would be set for life's chiral biochemistry to get going.
20. I discussed this fascinating story in *The Prophet and the Astronomer: Apocalyptic Science and the End of the World*. A detailed account can be found in Walter Alvarez, *T. Rex and the Crater of Doom* (Princeton, N.J.: Princeton University Press, 1997).
21. Readers can watch a short clip of the girl-to-gorilla transformation from the Bond movie in www.youtube.com/watch?v=KRlZvJhXefE. Younger readers may be familiar with the book series *Animorphs*, written by K. A. Applegate, which features adolescents with the power to transform into animals.
22. Perhaps surprisingly, the definition of what constitutes a species remains under debate. Although the ability to reproduce does demarcate a boundary between different species, complications exist. For example, coyotes and wolves interbreed, while microbes reproduce asexually. Given that microbes constitute approximately 90 percent of the tree of life, there is reason to think carefully about how to define a species. Some biologists go so far as to claim that the concept of a species is an illusion. Others think that ecological variables should be incorporated into the classification. For microbes, this would mean that those that adapt to live, say, at a certain water temperature and acidity are a different species from those living at different temperatures and acidities.
23. In reality the process is a bit more involved. DNA has two parts to it: a coding part and a noncoding part that sits between genes. Both are important.

The noncoding part selects which genes will be expressed at different times as animals evolve from egg and sperm to embryo to adulthood. This switching on and off of gene expression explains how such remarkable diversity of living creatures is achieved from essentially the same set of genes. These ideas form the basis of the recently proposed evolutionary developmental biology, or "evo devo." See, for example, Sean Carroll, *Endless Forms Most Beautiful* (New York: Norton, 2005).

PART V: THE ASYMMETRY OF EXISTENCE

1. All Einstein quotes in this book can be found in *The Quotable Einstein*, collected and edited by Alice Calaprice (Princeton, N.J.: Princeton University Press, 1996).
2. Paul Davies, *Cosmic Jackpot: Why Our Universe Is Just Right for Life* (New York: Houghton Mifflin, 2007). The U.K. edition of this book was titled *The Goldilocks Enigma: Why Our Universe Is Just Right for Life* (London: Penguin, 2006).
3. Alex Vilenkin, *Many Worlds in One: The Search for Other Universes* (New York: Hill & Wang, 2006), p. 143. Vilenkin offers a very lucid and thorough discussion of some of these issues and how they relate to the anthropic principle mentioned below.
4. For an accessible account see Max Tegmark's "Anything Goes," *New Scientist*, June 1998. Even Tegmark concedes this to be "one wacky [article]."
5. Johannes Kepler, letter to Jakob Bartsch, November 6, 1629.
6. See, for example, John Horgan, *The End of Science: Facing the Limits of Knowledge in the Twilight of the Scientific Age* (New York: Broadway, 1996). Horgan conveys his conviction that we may be nearing a stage in science where "further research may yield no more great revelations or revolutions, but only incremental, diminishing returns." How to reconcile such a notion with the fact that we should always be open to the unexpected is, to me, a mystery. New tools will invariably bring new challenges. It's simply impossible to predict that no grand changes or scientific revolutions will ever happen again. As history has shown over and over (think, for example, of the discovery of dark energy just two years after the publication of Horgan's book!), the laboratory and ever-evolving measuring tools are great destroyers of worldviews.
7. Beauty marks were all the rage in the European courts of the eighteenth century, and often since. If you didn't have one, you painted one on. For the curious reader I'd suggest the masterful movies *Amadeus*, directed by Milos Forman, and *Barry Lyndon*, directed by Stanley Kubrick. But as usual, what one culture finds beautiful another may find ugly. In Japan, beauty marks were frowned upon, being related to character flaws. In any case, the topic is very popular: in early July 2009, there were 79 million hits for "beauty mark" on Google.

8. For example, see Dahlia W. Zaidel and Choi Deblieck, "Attractiveness of Natural Faces Compared to Computer Constructed Perfectly Symmetrical Faces," *International Journal of Neuroscience* 117 (2007): 423–31.
9. A nice review of the fundamental constants, their deep significance to our description of Nature, and how they are measured can be found in Harald Fritzch's *Fundamental Constants in Physics: A Mystery of Physics* (Singapore: World Scientific, 2008). A more technical discussion can be found in Max Tegmark, Anthony Aguirre, Martin Rees, and Frank Wilczek, "Dimensionless Constants, Cosmology, and Other Dark Matters," *Physical Review* D 73 (2006): 023505.
10. For the purists, it also depends on the permittivity of space, a constant originating in electromagnetism and related to the speed of light.
11. In 1999, Lisa Randall and Raman Sundrum proposed a speculative theory, known as "brane theory," which relies on large, as opposed to small, extra dimensions. Details can be found in Randall's popular account, *Warped Passages: Unraveling the Mysteries of the Universe's Hidden Dimensions* (New York: HarperCollins, 2005). I will consider only the more "conventional" small extra dimensions here.
12. A wonderful exploration of our need to believe can be found in Carl Sagan's *The Demon-Haunted World: Science as a Candle in the Dark* (New York: Ballantine, 1996). I should also mention that in *Cosmic Landscape,* Leonard Susskind uses a "clever fish" metaphor to argue for the anthropic principle. Amusingly, I only found this out after writing the lines above. I guess our fish are clever (and foolish) in different ways.
13. Olaf Stapledon, *The Star Maker* (Middletown, Conn.: Wesleyan University Press, 2004), p. 234. This is a nice edition of the 1937 classic, with a foreword by Freeman Dyson.
14. A wonderful account of Lynn Margulis's work on endosymbiosis can be found in the book she co-authored with her son Dorion Sagan, *Microcosmos: Four Billion Years of Microbial Evolution* (Berkeley: University of California Press, 1997).
15. This quote and a lively telling of the discovery of ALH84001 can be found in Paul Davies, *The Fifth Miracle: The Search for the Origin and Meaning of Life* (New York: Simon & Schuster, 1999). Davies also offers a fascinating discussion of the science and philosophical implications of discovering alien life.
16. Remarkably, Kepler wrote what is probably the first modern account of life elsewhere. His short story "The Dream" was published posthumously in 1634 after much inner turmoil. With uncanny foresight into natural selection, Kepler described bizarre lunar creatures, adapted to an environment very different from Earth's. His main intention, though, was to describe astronomy from the perspective of an orbiting celestial body as evidence that Earth too could be in orbit.
17. Arthur C. Clarke, *2001: A Space Odyssey* (New York: New American Library, 1968), p. 245.

18. Davies, *The Fifth Miracle*, p. 246.
19. Stephen Webb wrote a very amusing book, exploring fifty different answers to this question: *If the Universe Is Teeming with Aliens . . . Where Is Everybody? Fifty Solutions to the Fermi Paradox and the Problem of Extraterrestrial Life* (New York: Copernicus, 2002).
20. Many books and articles have been written on this topic and I list a few in the bibliography. I also direct the interested reader to the website www.actionbioscience.org, run by the American Institute of Biological Sciences. It is an excellent resource for more references, comprehensive articles, and websites.
21. *The Prophet and the Astronomer: Apocalyptic Science and the End of the World* (New York: Norton, 2003).

BIBLIOGRAPHY

Adams, Fred, *Our Living Multiverse: A Book of Genesis in 0 + 7 Chapters*. New York, NY: Pi Press, 2003.

Adams, Fred, and Laughlin, Greg, *The Five Ages of the Universe*. New York, NY: The Free Press, 1999.

Adler, Mortimer J. (ed.), *Great Books of the Western World*. Chicago, IL: Encyclopedia Britannica, 1990.

Alvarez, W., *T. Rex and the Crater of Doom*. Princeton, NJ: Princeton University Press, 1997.

Armstrong, Karen, *A History of God*. New York, NY: A. A. Knopf, 1993.

Barrow, John D., *Between Inner Space and Outer Space: Essays on Science, Art, and Philosophy*. Oxford, UK: Oxford University Press, 1999.

Barrow, John D., and Silk, Joseph, *The Left Hand of Creation: The Origin and Evolution of the Expanding Universe*. New York, NY: Basic Books, 1983.

Barrow, John D., and Tipler, Frank J., *The Anthropic Cosmological Principle*. New York, NY: Oxford University Press, 1996.

Berlin, Isaiah, *Concepts and Categories*. New York, NY: Viking Press, 1979.

Boorstin, Daniel J., *The Discoverers*. New York, NY: Vintage, 1985.

Burkert, Walter, *Lore and Science in Ancient Pythagorianism*, trans. Edwin L. Milnar, Jr. Cambridge, MA: Harvard University Press, 1972.

Carroll, Lewis, *Alice in Wonderland and Through the Looking Glass*. New York, NY: Grosset & Dunlap, 1946.

Carroll, Sean, *Endless Forms Most Beautiful*. New York, NY: W. W. Norton, 2005.

Clarke, Arthur C., *2001: A Space Odyssey*. New York, NY: New American Library, 1968.

Cole, K. C., *The Hole in the Universe*. New York, NY: Harcourt, 2001.

Crick, Francis, *Life Itself: Its Nature and Origin*. New York, NY: Simon & Schuster, 1981.

Davies, Paul, *The Mind of God*. New York, NY: Simon & Schuster, 1992.

Davies, Paul, *About Time*. New York, NY: Simon & Schuster, 1995.

Davies, Paul, *Are We Alone?* New York, NY: Basic Books, 1995.

Davies, Paul, *The Fifth Miracle: The Search for the Origin and Meaning of Life*. New York, NY: Simon and Schuster, 1999.

Davies, Paul, *Cosmic Jackpot: Why Our Universe is Just Right for Life*. New York, NY: Houghton Mifflin, 2007.

Dawkins, Richard, *The God Delusion*. New York, NY: Houghton Mifflin, 2006.

Dawkins, Richard, *The Greatest Show on Earth: The Evidence for Evolution*. New York, NY: Free Press, 2009.

De Fontenelle, Bernard le Bovier, *Conversations on the Plurality of Worlds* (1687). Berkeley, CA: University of California Press, 1990.

Dyson, Freeman, *Origins of Life* (1985). Princeton, NJ: Princeton University Press, 1999 (2nd ed.)

Dyson, Freeman, *The Sun, the Genome, and the Internet*. New York, NY: Oxford University Press, 1999.

Einstein, Albert, *The Quotable Einstein*, collected and edited by Alice Calaprice. Princeton, NJ: Princeton University Press, 1996.

Fenchel, Tom, *Origin and Early Evolution of Life*. Oxford, UK: Oxford University Press, 2002.

Frank, Adam, *The Constant Fire: Beyond the Science vs. Religion Debate*. Berkeley, CA: University of California Press, 2009.

Fritzch, Harald, *Fundamental Constants in Physics: A Mystery of Physics*. Singapore: World Scientific, 2008.

Gingerich, Owen, *The Book Nobody Read*. New York, NY: Walker & Sons, 2004.

Gleiser, Marcelo, *The Dancing Universe: From Creation Myths to the Big Bang*. New York, NY: Plume, 1998. [Hardcover edition: Dutton, 1997.]

Gleiser, Marcelo, *The Prophet and the Astronomer: A Scientific Journey to the End of Time*. New York, NY: W. W. Norton, 2002.

Goodenough, Ursula, *The Sacred Depths of Nature*. Oxford, UK: Oxford University Press, 1998.

Gould, Stephen J., *Dinosaur in a Haystack*. New York, NY: Harmony Books, 1995.

Greene, Brian, *The Elegant Universe: Superstrings, Hidden Dimensions, and the Quest for the Ultimate Theory*. New York, NY: W. W. Norton, 1999.

Grinspoon, David, *Lonely Planets: The Natural Philosophy of Alien Life*. New York, NY: Ecco, 2003.

Guth, Alan, *The Inflationary Universe: The Quest for a New Theory of Cosmic Origins*. Reading, MA: Addison-Wesley, 1997.

Hawking, Stephen, *A Brief History of Time: From the Big Bang to Black Holes*. New York, NY: Bantam Books, 1988.

Holton, Gerald, "Einstein and the Goal of Science," in *Einstein, History, and Other Passions*. Cambridge, MA: Harvard University Press, 1996.

Hooper, Dan, *Nature's Blueprint: Supersymmetry and the Search for a Unified Theory of Matter and Force*. New York, NY: HarperCollins, 2008.

Horgan, John, *The End of Science: Facing the Limits of Knowledge in the Twilight of the Scientific Age*. New York, NY: Broadway Books, 1996.

Kahn, Charles, *Pythagoras and the Pythagoreans: A Brief History*. Indianapolis, IN: Hackett, 2002.

Kauffman, Stuart, *At Home in the Universe*. Oxford, UK: Oxford University Press, 1995.

Kauffman, Stuart, *Reinventing the Sacred: A New View of Science, Reason, and Religion*. New York, NY: Basic Books, 2008.

Kepler, Johannes, *Mysterium Cosmographicum* (1596). New York, NY: Abaris Books, 1981.

Kirk, G. S., Raven, J. E., *The Presocratic Philosophers*. Cambridge: Cambridge University Press, 1971.

Kolb, Rocky, *Blind Watchers of the Sky: The People and Ideas that Shaped our View of the Universe*. New York, NY: Basic Books, 1997.

Krauss, Lawrence, *Quintessence: The Mystery of Missing Mass in the Universe*. New York, NY: Basic Books, 2000.

Lindley, David, *The End of Physics: The Myth of a Unified Theory*. New York, NY: Basic Books, 1993.

Livio, Mario, *The Accelerating Universe*. New York, NY: John Wiley & Sons, 2000.

Livio, Mario. *Is God a Mathematician?* New York, NY: Simon & Schuster, 2009.

Margulis, Lynn, and Sagan, Dorion, *Microscosmos: Four Billion Years of Microbial Evolution*. Berkeley, CA: University of California Press, 1997.

Mather, John C., & Boslough, J., *The Very First Light: The True Inside Story of the Scientific Journey Back to the Dawn of the Universe*. New York, NY: Basic Books, 1996.

Monod, Jacques, *Chance and Necessity*, trans. A. Wainhouse. London, UK: Collins, 1972.

Munitz, Milton K. (ed.), *Theories of the Universe: from Babylonian Myth to Modern Science*. Glencoe, IL: Free Press, 1957.

North, John, *The Norton History of Astronomy and Cosmology*. New York, NY: W. W. Norton, 1995.

Orgel, Leslie, *The Origins of Life: Molecules and Natural Selection*. New York, NY: John Wiley & Sons, 1973.

Pasteur, Louis, *Researches on the Molecular Asymmetry of Natural Organic Products*. Edinburgh, UK: The Alembic Club, 1905.

Randall, Lisa, *Warped Passages: Unraveling the Mysteries of the Universe's Hidden Dimensions*. New York, NY: HarperCollins, 2005.

Rees, Martin, *Before the Beginning: Our Universe and Others*. New York, NY: Perseus Books, 1997.

Rees, Martin, *Our Cosmic Habitat*. Princeton NJ: Princeton University Press, 2001.

Rees, Martin, *Our Final Hour: A Scientist's Warning How Terror, Error, and Environmental Disaster Threatens Humankind's Future in This Century—On Earth and Beyond*. New York, NY: Basic Books, 2003.

Roach, Mary, *Stiff: The Curious Lives of Human Cadavers*. New York, NY: W. W. Norton, 2003.

Sagan, Carl, *Pale Blue Dot: A Vision of the Human Future in Space*. New York, NY: Ballantine, 1994.

Sagan, Carl, *The Demon-Haunted World: Science as a Candle in the Dark*. New York, NY: Ballantine, 1997.

Sagan, Carl, *The Varieties of Scientific Experience: A Personal View of the Search for God*. New York, NY: Penguin Press, 2006.

Schrödinger, Erwin, *What Is Life?* (1958). Cambridge, UK: Cambridge University Press (Canto Edition), 1992.

Smolin, Lee, *Three Roads to Quantum Gravity*. New York, NY: Basic Books, 2001.

Smolin, Lee, *The Trouble with Physics: The Rise of String Theory, the Fall of a Science, and What Comes Next*. New York, NY: Houghton Mifflin Harcourt, 2006.

Smoot, George, and Davidson, Keay. *Wrinkles in Time*. New York, NY: W. Morrow, 1993.

Stapledon, Olaf, *The Star Maker*, Middletown, CT: Wesleyan University Press, 2004.

Stewart, Ian, *Why Beauty is Truth: A History of Symmetry*. New York, NY: Basic Books, 2007.

Sullivan, Woodruff T. III, and Baross, John A. (eds.), *Planets and Life: The Emerging Science of Astrobiology*. Cambridge, UK: Cambridge University Press, 2007.

Susskind, Leonard, *The Cosmic Landscape: String Theory and the Illusion of Intelligent Design*. New York, NY: Little, Brown and Company, 2006.

Vilenkin, Alex, *Many Worlds in One: The Search for Other Universes*. New York, NY: Hill & Wang, 2008.

Webb, Stephen, *Where is Everybody? Fifty Solutions to the Fermi Paradox and the Problem of Extraterrestrial Life*. New York, NY: Copernicus Books, 2002.

Weinberg, Steven, *The First Three Minutes: A Modern View of the Origin of the Universe* (1979). New York, NY: Basic Books, 1993 (2nd edition).

Weinberg, Steven, *Dreams of a Final Theory: The Search for the Fundamental Laws of Nature*. New York, NY: Pantheon Books, 1993.

Weinstein, Steven, "Anthropic Reasoning and Typicality in Multiverse Cosmology and String Theory." *Classical and Quantum Gravity* 23 (2006) 4231–36.

Wilczek, Frank, *The Lightness of Being: Mass, Ether, and the Unification of Forces*. New York, NY: Basic Books, 2008.

Wilczek, Frank, and Devine, Betsy, *Longing for the Harmonies: Themes and Variations from Modern Physics*. New York, NY: W. W. Norton, 1988.

Willis, Christopher, and Bada, Jeffrey, *The Spark of Life: Darwin and the Primeval Soup*. New York, NY: Perseus Books, 2000.

Woit, Peter, *Not Even Wrong: The Failure of String Theory and the Continuing Challenge to Unify the Laws of Physics*. London, UK: Jonathan Cape, 2006.

ACKNOWLEDGMENTS

I am fortunate to have so many colleagues and friends who were willing to take time out of their busy schedules to read the manuscript and offer their invaluable advice and suggestions. If errors and omissions persist, they are entirely my doing. First, I should thank Agnes Krup, who helped an idea become a book and, once it did, made sure many would have access to it. My wonderful agent Michael Carlisle and his assistant Ethan Bassoff at Inkwell did an amazing job keeping me on track, offering criticism and inspiring suggestions along the way. Hilary Redmon, my editor at Free Press, deserves all the praise in the world: her enthusiastic support and ideas made for a much better book. I thank Richard Kremer, Mark McPeek, and Adam Frank for their valuable input and, especially, Nancy Frankenberry, my brother Luiz, Sara Walker, and Steve Weinstein for reading the complete manuscript and for their excellent suggestions. Nicole Younger-Halpern, my Presidential Scholar at Dartmouth, read and critiqued the complete manuscript as a true professional. Finally, I thank my wife Kari for finding the time and energy to read the manuscript and helping me improve it, even after long days helping others at her office. A last word of thanks goes to my brothers, for being a constant source of inspiration and support in my life.

INDEX

Abiogenesis, 165, 188
Absurd Universe, 217–218, 222
Accelerating Universe, 95–97, 127, 150
Accidental Universe, 217–218, 222
Aces, 118
Adam and Eve, 4
Adenine, 209
Adenosine, 185
Aether, 47–48, 67, 125
Afterlife, belief in, 19
Agnosticism, 18
Akhenaten, 16, 20–21
Akilia Island, 169
Aldrin, Buzz, 12
ALH84001, 239–240
Alien life, xvi, 13, 200, 203–204, 212, 238–247
Alpha Centauri, 72, 246
Alpher, Ralph, 57
Alternating current, 49
Amadeus (Shaffer), 29
Amino acids, 165, 166, 180, 182, 185, 193, 194, 199–204, 209, 236
Ammonia, 164, 165, 174, 174*n*, 235
Anaximander, 103
Anaximenes, 22
Anderson, Carl, 111, 121
Anderson, Philip, 177
Andromeda, 72
Antarctica, 239
Anthropic principle, 127, 220–221, 233
Anthropocentric view, 249
Antibaryons, 135–136, 140
Antihydrogen, 114
Antimatter, 113–116, 121, 129, 135, 136, 138, 140, 143, 145, 147, 199, 212, 229
Antineutrinos, 107, 108, 130–133
Antiparticles, 113, 114
Antiprotons, 113
Antiquarks, 135
Arcadia (Stoppard), 6
Aristarchus of Samos, 26
Aristotle, 6, 22, 25–27, 32, 96, 187
Armstrong, Neil, 12, 91, 114
Art, 102
Asimov, Isaac, 14
Asteroids, 205, 211, 239*n*
Astronomia Nova (*New Astronomy*) (Kepler), 31, 37, 59
Asymptotic freedom, 120
Atheism, 18
Atomic number, 110*n*, 115–116, 141
Atoms, 15, 52, 62, 65, 84, 109–110, 173, 174*n*, 177, 177*n*, 192
  birth of, 55–57
Australia, 169, 180, 181, 204
Autocatalytic reactions, 196–199
Avogadro number, 115*n*, 137*n*
Awe, search for meaning and, 222–223
Axions, 134

B mesons, 134
Babylonians, 25
Bacteria, 173, 182, 187–188, 240
Baryogenesis, 136, 137, 139, 140, 144
Baryons, 96*n*, 118–120, 135–136, 140, 142–145

Belief, power in, 16–19
Berg, Alban, 102–103
Berlin, Isaiah, 23, 123
Beta decay, 130–132
Big Bang model, 5, 43, 46, 56, 57, 65, 66, 71, 74, 76, 80, 88–90, 97, 115, 125, 136, 142, 149, 156, 224, 225
  faltering, 78–79
Biology, 174
Biomolecules, 200–201, 235
Biot, Jean-Baptiste, 191–192
Black holes, 15, 93, 114
Blake, William, 46
Blue-green algae, 182, 183
Bohr, Niels, 130
Bombardment era, 167–169
Book of Genesis, 4, 5, 58
Borges, Jorge Luis, 233
Boson stars, 93
Botanic Garden, The (E. Darwin), 176
Brahe, Tycho, 31, 37, 103
Brahman, 20*n*
Brain hemispheres, 228
Brandenburg, Axel, 202
Bronowski, Jacob, 224
Brownlee, Donald, 238, 240
Buddha, 20*n*
Buddhism, 20*n*
Byron, Lord, 161

Cabrera, Blas, 53
Calloway County, Kentucky, 204*n*
Caloric, 48
Cambrian explosion, 183*n*, 237
Carbohydrates, 182
Carbon, 64–65, 157, 169, 170, 173, 174*n*, 235
Carbon dioxide, 165, 235, 237
Carneiro, Gilson, 45, 52, 163
Carter, Brandon, 220
Catastrophism, 208
Catholic Church, 29
Celestial Mechanics (Laplace), 60–61
Cells, 179, 182, 185, 209
Chadwick, James, 110, 111, 129–130
Chaney, Lon, Jr., 206
Charge conjugation, 129, 132, 147
  CP violation, 133–137, 139, 143, 144
Chariots of the Gods (Däniken), 13
Chirality, 193, 194–205
Chlorine, 229–230
Chlorophyll, 238
Cicero, 223
Clarke, Arthur C., 241
Clays, 175
Climate changes, 237, 250
Cline, David, 115
Clinton, Bill, 239–240
Close Encounters of the Third Kind (movie), 247
COBE (Cosmic Background Explorer) satellite, 74, 75
Cold War, 244
Colonial Theory, 237
Columbus, Christopher, 91, 96
Comets, 157, 239*n*
Conservation laws, 111, 112
Contact (Sagan), vii, 247
Conversations on the Plurality of Worlds (De Fontanelle), 216
Copernicus, Nicolaus, 25–29, 31, 32, 43, 59
Core of Earth, 52*n*
Cosmic Jackpot: Why Our Universe is Just Right for Life (Davies), 217, 220
Cosmological inflation, 81–85, 87–89, 95, 97, 139, 148, 156, 180, 186
Crawford, Cindy, 228, 229
Creation myths, 58–59
Cretaceous era, 205, 208
Crick, Francis, 196
Critical energy density, 76
Critical point, 203
Cronin, James, 133, 135, 204

Crust of Earth, 175
Cultural relativism, 226*n*
Cyanobacteria, 169
Cytidine, 185
Cytosine, 209

Däniken, Erich von, 13
Dante Aligheri, 32
Dark Ages, 25
Dark energy, 95–97, 104, 127, 128, 221, 227, 229, 232
Dark matter, 55*n*, 92–97, 138, 148, 156, 229
Darwin, Charles, 82, 171, 172, 176, 179, 206–209
Darwin, Erasmus, 161, 176
Darwin mission, 238
Davies, Paul, 217, 220, 242
Dawkins, Richard, 17, 48, 224
Day the Earth Stood Still, The (movie), 244
De Fontenelle, Bernard le Bovier, xiii, 216
Death, 161, 163–164
Deists, 60
Demiurge, 27
Democritus, 109
Dennett, Daniel, 17
Determinism, 60–61, 242
Deuterium, 55, 110
Devine, Betsy, 19
Dharmakaya, 20*n*
Diamond, 170
Diamonds Are Forever (movie), 206
Dinosaurs, 205, 208, 211, 212, 243, 250
Diogenes Laërtius, 22
Dirac, Paul Adrian Maurice, 53, 111, 113, 130
Divine Comedy (Dante), 32
DNA, 182, 184–185, 193, 194, 196, 208–211, 237
Drake, Frank, 243
Drake equation, 243–244
Dreams of a Final Theory: The Search for the Fundamental Laws of Nature (Weinberg), 19, 104, 123
Dukas, Paul, 196
Dyson, Freeman, 183, 185

Earth-centered model, 25–27, 32, 59, 96
Eigen, Manfred, 184
Einstein, Albert, xiv, 14, 19, 47, 48, 61–62, 75, 83, 97, 124, 125, 127, 147, 149, 217, 225
  general theory of relativity of, 52, 67–69, 84*n*, 91–93, 104
  special theory of relativity of, 6, 15, 51–52, 67–68, 111, 113, 134
Elastic potential energy, 84
Eldredge, Niles, 205
Electricity, 49–50, 61, 85, 117, 159–161, 164
Electromagnetism, 45, 47–50, 54–55, 61, 65, 87, 96*n*, 113, 117, 121, 125, 136, 139, 141, 147, 148
  imperfection of, 51–53
Electron-positron pair, 112
Electrons, 52, 53, 55–57, 62, 65, 69, 70, 73, 84, 86, 88, 89, 94, 96, 107, 109–113, 121, 122, 135, 148, 177*n*, 229–231
Electroweak phase transition, 142, 144, 145, 148
Elementary particles, 55
Energy conservation, 130, 147
Energy density, 83–85
Enlightenment, 60
Enzymes, 195
Epicycles, 26, 26*n*
Equilibration time, 79
Equilibrium temperature, 79
Euclidean geometry, 75
Eukaryotes, 182, 183, 235–237
Europa, xvi, 236
Evolution, theory of, 82, 156, 207, 242

Evolution of Physics, The (Infeld), 14
Evolutionary convergence, 211
Expansion of Universe, 56–57, 136, 156
External symmetries, 106–108, 129, 141, 142
Extinctions, 237, 250
Extra-dimensional space-time, 5
Extraterrestrial intelligence, xvi, 13, 200, 203–204, 212, 238–247
Extremism, 17

Faith, 46, 216–217
Fantasia (movie), 196–197
Faraday, Michael, xiv, 49–51
Fenchel, Tom, 184, 185
Fermentation, 187
Fermi, Enrico, 244
Fermi's Paradox, 244
Final Theory, xiv, 19, 124–126, 224–226
Final Truth, 6, 7, 69, 70, 87, 89, 126, 127, 147, 149, 150, 222, 223, 226
Finnegan's Wake (Joyce), 119
First Cause, 3–5, 7
First life. *See* Life, origins of
First Three Minutes, The: A Modern View of the Origin of the Universe (Weinberg), 45, 46, 55, 64, 80, 163
Fitch, Val, 133, 135
Flatness problem, 79, 81, 89
Ford, W. K., 92
Fossils, 167, 207–208, 211
"Four Horsemen," 17
Frank, Sir Frederick Charles, 196, 197, 199, 200, 202
Frankenstein: Or, the Modern Prometheus (Shelley), 13, 161–164
Franklin, Benjamin, 60, 91, 159, 160
Freud, Sigmund, 21

Galapagos Islands, 209
Galaxies, 55, 55*n*, 56, 66, 72, 73, 75, 76, 91–95, 186, 212, 249
Galileo Galilei, 59, 60
Galvani, Luigi, 159–160
Galvanism, 160, 161
Gamma-ray radiation, 112, 116
Gamow, George, 14, 56–57, 64
Gao, Y., 115
Garriga, Jaime, 221
Gell-Mann, Murray, 118–120
General theory of relativity, 52, 67–69, 75, 83, 84*n*, 91–93, 104
Genesis, Book of, 4, 5, 58
Genetics, 6, 182–185, 208–210
Genome, 210, 237
Genotype, 208
Genus Homo, 237
Geometry, 33, 35, 37, 75, 76
Georgi, Howard, 137
Giraffes, 210
Glashow, Sheldon, 121, 137, 140
Gleiser, Luiz, 171, 172
Gluons, 116, 119–121, 136
God, belief in, 16, 60
Goldin, Daniel S., 239
Gould, Stephen Jay, 205
Gradualism, 206–208
Grand Unified Theories (GUTs), 87–90, 136–137, 139, 147–149
Graphite, 170
Gravitation, universal law of, 6
Gravitational lensing, 93
Gravity, 37, 50, 64, 67–69, 87, 96, 96*n*, 117, 125
Great Hymn to Aten (Akhenaten), 20
Greenland, 169
Gross, David, 120, 219
Guanine, 209
Guth, Alan, 80–81, 87, 88

Hadrons, 118–121, 131
Handedness, 101–102, 132–134, 148, 193–197, 200–204

Harmonic law, 37*n*
Harmonic sounds, 24
Harris, Sam, 17
Hawking, Stephen, 7
Heisenberg, Werner, xiv, 6, 149
Heisenberg Uncertainty Principle, 62
Helium, 73, 110, 131, 156, 157
Helium-3, 55
Helium-4, 55
Hemispheres of brain, 228
Heraclides of Pontus, 25–26
Hermann, Robert, 57
Higgs, Peter, 86
Higgs particle, 86, 131, 138, 140–142, 144–145, 148, 229
Hinduism, 20*n*
His Dark Materials trilogy (Pullman), 93
History and Present Status of Electricity (Priestly), 159
History of Western Philosophy (Russell), 23
Hitchens, Christopher, 17
Holton, Gerald, 22, 37
Horizon, 76–77
Horizon problem, 79, 81, 89, 136
Horror movies, 206, 207
Hoyle, Fred, 64
Hubble, Edwin, 56, 66, 76, 91
Hubble Space Telescope, 76
Hydrogen, 64–65, 92, 110, 115, 131, 141, 156, 157, 164, 229, 235, 242, 248–249
Hydrothermal vents, 175

Induced fit model, 195*n*
Inertia, 86
Infeld, Leonard, 14
Inflationary theory, 81–85, 87–89, 95, 97, 139, 148, 156, 180, 186
Integer numbers, 24
Internal symmetries, 106–108, 110, 129–131, 140, 142, 147
Ionian Delusion, 123
Ionian Enchantment, 22–24, 38, 118, 123, 150
Ionian Fallacy, 23, 123
Iron, 64, 65*n*, 157
Isotopes, 110
Isovaline, 204
Itacuruçá, island of, 171–172

Jefferson, Thomas, 60
Joyce, Gerald, 184
Joyce, James, 119
Jupiter, 26, 157
"Just right" for life theory, 231–234, 238, 240

Kant, Immanuel, 124
Karloff, Boris, 162
Kasting, James, 238
Kauffman, Stuart, 177
Keats, John, xiv, 104
Kepler, Johannes, xiv, 6, 14, 19, 28–39, 43, 59, 60, 103, 124, 147, 149, 150, 217, 222–223
Kepler mission, 238
Kolb, Rocky, 137
Kondepudi, Dilip, 202
Kubrick, Stanley, 13, 244
Kuiper belt, 157
Kuzmin, Vadim, 142–143

Lancet, Doron, 184
Landscape, 70
Laplace, Pierre Simon de, 60–62, 106
Large Hadron Collider (LHC), 86, 131, 138, 148
Last universal common ancestor (LUCA), 209, 211
Late heavy bombardment, 168*n*, 170
Lawrence Berkeley National Laboratory, 95
Lee, T. D., 132
Left-handed people, 101–102
Leptons, 96*n*, 116, 121–122, 123, 131, 132, 135, 137

Leucippus, 109
Leyden jars, 159, 160
Library of Babel (Borges), 233
Life, definition of, 172–173
Life, origins of, xvii, 155–158, 165
  building blocks, 182–186
  chirality, 193, 194–205
  "how" question, 173, 177–181
  Pasteur on, 187–195
  spontaneous generation view, 187–190
  "when" question, 167–170
  "where" question, 171–176, 181, 201
Life principle, 220, 222
Light, 47–49, 51–52, 54, 61–62, 70, 72, 73, 80–81
Light nuclei, 80, 84
Light-year, 72*n*
Lightness of Being: Mass, Ether, and the Unification of Forces (Wilczek), 126*n*
Lightning, 164
Linde, Andrei, 221
Lindley, David, 124
Lithium, 64–65, 73
Lithium-7, 55, 115
Lock-and-key model, 195, 195*n*
Loew, Rabbi, 11
"Logical Translation" (Berlin), 23
Longing for the Harmonies: Themes and Variations from Modern Physics (Wilczek and Devine), 19
Loop quantum gravity, 69
Luminiferous aether, 47–48, 67, 125
Luther, Martin, 29
Lutheran Church, 29
Lyell, Charles, 206–207

MacDougall, Duncan, 163
Maestlin, Michael, 28–32, 35
Magnetic field, 238, 243–244
Magnetic monopoles, 52–53, 113, 129, 147
Magnetism, 49, 52, 129
Mahayana Buddhism, 20*n*
Mammals, 205, 211
Manhattan Project, 111
Margulis, Lynn, 237
Mars, xvi, 26, 59, 103, 157, 201, 236, 238–239
Marx, Karl, 18, 222
Mass, 124, 140, 141
Mass extinctions, 237, 250
Mathematical Principles of Natural Philosophy (Newton), 60, 216
Mathematics, 23–24, 88
Mather, John, 75
Matrix, The (movie), 244
Matter, 70, 75, 80, 81, 83–86, 91–94, 110–120, 129–134, 140–147, 156, 199, 201–202, 229
  origin of, 135–139
Maxwell, James Clerk, 51, 113, 129
Mayr, Ernst, 210
McCarthy, Cormac, 245
McKay, Christopher, 239
McMurdo Dry Valleys, 239
Mediocrity, principle of, 219, 242
Megaverse, 4
Meiosis, 208
Mendeleyev, Dmitry, 109, 110*n*, 119
Mercury, 26, 157, 167, 238
Mesons, 118–120
Metabolism, 183, 185
Meteorites, 165, 180–181, 204, 205, 208, 239–240
Methane, 164, 165, 235, 238
Michelson, Albert, 48, 125
Microwaves, 45, 47, 55, 73–75, 78, 136, 156
Milky Way, 72, 73, 156, 243
Miller, Stanley, 164–165
  Miller-Urey spark-of-life experiments, 164–166, 180, 185, 195, 196, 199, 236
Mitochondria, 237
Modern Synthesis, 208

Molecules, 65, 155, 165–166, 168, 173, 174, 177, 180–182, 184, 186, 191–197, 199–203, 212, 223, 235, 236, 249
Momentum, 107, 107*n*, 112
Monotheism, xiv, 20–21, 69, 123, 222
Monroe, Marilyn, xv, 228, 229
Moon, 27, 59, 76, 91, 114, 167–168, 175, 175*n*
Morley, Edward, 48
Moses and Monotheism (Freud), 21
Motion, third law of, 92*n*
Mount Stromlo Observatory, 95
Mozart, Wolfgang Amadeus, 29
Multamäki, Tuomas, 202
Multi-brane collision, 5
Multicellular organisms, 183, 183*n*, 235, 237, 238, 242–243
Multiverse theories, 4, 219–222, 231
Muon, 121
Murchison meteorite, 180, 181, 204
Murray meteorite, 204, 204*n*
Music, 102–103
Muslim astronomers, 25
Mutations, 208–212, 223
Mutual antagonism, 197
Mysterium Cosmographicum, The (Kepler), 31, 34–37

Napoleon, Emperor, 60–61
NASA (National Aeronautics and Space Administration), 74, 238, 239
Natural selection, 156, 184, 185, 200, 207, 210, 211
Naturalists, 216, 224
Nature, hidden code of, xiii–xvi, 6–7, 14, 15, 19, 23–24, 36, 60, 106, 124, 218, 223
Naumann, Robert, 226
Nazca Lines, Peru, 13
Needham, John, 188
Neoplatonists, 23
Neptune, 91, 157
Neutral kaon, 133
Neutrinos, 55, 84, 88, 91, 92, 96*n*, 110*n*, 111, 113, 117–119, 121, 122, 129–133, 140, 148, 157, 193, 200, 232
Neutrons, 55, 64*n*, 65, 80, 84, 89, 96*n*, 107, 108, 110, 130, 131, 135, 231
Newton, Isaac, xiv, 6, 37, 38, 60, 82, 216
Nitrogen, 64, 157, 165, 235
Nordita, 197, 202
Not Even Wrong: The Failure of String Theory and the Continuing Challenge to Unify the Laws of Physics (Woit), 125
Nuclear fusion, 64, 157
Nuclear physics, 64
Nuclei, 64–65, 130
Nucleic acids, 196, 236
Nucleosides, 185

Ockham's razor, 82, 91, 137, 225
Octet classification, 118, 119
On the Origin of Species (C. Darwin), 179, 207
On the Revolutions of the Heavenly Spheres (Copernicus), 25, 27–28
Oort cloud, 157
Oparin, Alexander, 183–184
Oppenheimer, J. Robert, 111
Organic compounds, 165
Orgel, Leslie, 168–169, 184
Origin and Early Evolution of Life (Fenchel), 184, 185
Origin of Life, The (Oparin), 183
Origins of Life (Dyson), 185
Osiander, Andreas, 28
Out-of-equilibrium process, 143–144, 178
Oxygen, 64, 141, 157, 169, 176, 183, 235, 237, 238
Ozone, 183, 238

Panspermia, 166
Paris Academy of Sciences, 188
Parity, 131–132, 147
CP violation, 133–137, 139, 143, 144, 200
Particle physics, 46, 65, 66, 103, 107–109, 114, 116, 124, 128, 129, 147, 199
Standard Model of, 85–87, 93, 104, 123, 125, 134, 139, 140, 143, 145–146, 148, 225
Pasteur, Louis, vii, 187–195, 200–202
Paul III, Pope, 28
Pauli, Wolfgang, 130, 131, 149
Pella Migdol Temple, Jordan, 21
Penn State University, 73, 238
Penzias, Arno, 43, 57, 64, 74
Peptide nucleic acids (PNAs), 195
Peptides, 185
Periodic table, 109, 110, 119
Perlmutter, Saul, 95
Pew Forum for Religion and Public Life poll (2008), 16, 17
Phase transitions, 85, 140–145
Phenotype, 208
Philosophy, 21, 22, 82
Phlogiston, 48
Photons, 62, 66, 69, 73, 74, 84, 88, 91, 96*n*, 112, 121, 140, 141, 145, 156, 179*n*
Photosynthesis, 169
Picasso, Pablo, 102
Pions, 118, 119
Pizzarello, Sandra, 204
Planck, Max, 6
Planetesimals, 157, 167
Planets, 26–27, 32, 33, 35–38, 59, 60, 91–92, 94, 124, 157, 158, 167, 212, 236, 238–239, 243–244
Plank's constant *h*, 229
Plasson, Raphaël, 197
Plate tectonics, 237, 243
Plato, xiv, 23, 27, 36, 105
Plotinus, 23
Pluto, 157
Plutonium, 65
Polarized light, 190–193
Politzer, David, 120
Polymerization process, 184, 196
Positrons, 111–113
Potential energy, 84
Pressure, 83–85, 84*n*, 95
Priestly, Joseph, 159
Primordial nucleosynthesis, 80, 115
Principia. See Mathematical Principles of Natural Philosophy (Newton)
Problem with Physics, The: The Rise of String Theory, the Fall of a Science, and What Comes Next (Smolin), 125
Prokaryotes, 174, 182, 183, 235–237
Protagoras, xv
Proteins, 165, 173, 193, 194, 196, 209, 212, 236
Protocells, 174, 184, 185
Protons, 55–57, 64*n*, 65, 66, 73, 80, 84, 86, 89, 94, 96, 96*n*, 107, 108, 110*n*, 110–113, 116, 118, 119, 122, 124, 131, 135, 137, 143, 145, 147, 148
Protozoa, 236, 239
Ptolemy, 25–27, 26*n*
Pullman, Philip, 93
Punctuated equilibrium hypothesis, 205
Pythagoras, xiv, 6, 23–24, 36
Pythagoreans, 23–24, 30, 33

Quantum chromodynamics (QCD), 119
Quantum theory, 53, 62–63, 68–70, 96, 98, 111, 113, 114, 127, 177, 226
Quantum vacuum decay, 5
Quarks, 116, 118–121, 123, 132, 135–137, 229
Quintessence, models of, 96

Radiation, 43, 54–57, 73, 74, 78–81, 83–84, 112, 147, 157, 183, 201, 211, 238
Radical atheism, 17
Radioactivity, 55, 110, 121, 130
Ramos, Rudnei, 145
Rare Earth: Why Complex Life Is Uncommon in the Universe (Ward and Brownlee), 238, 240, 243
Reason, 216–217, 248
Redi, Francesco, 187, 188
Reductionism, 123, 124, 177
Relativity
 general theory of, 52, 67–69, 75, 83, 84*n*, 91–93, 104
 special theory of, 6, 15, 51–52, 67–68, 111, 113, 134
Religion
 faith, 48, 216–217
Religious belief, xiii–xiv, 4, 13–14
 enduring power of, 17–18
 incidence of, 16–17
Renaissance, 23, 24, 26
Riess, Adam, 95
RNA, 184–185, 193, 194
Road, The (McCarthy), 245
Rocks, 167–170, 207
Romantics, 60, 163
Rotation of Earth, 175*n*
Rotational symmetry, 106
Rubakov, Valery, 142–143
Rubin, Vera, 92
Russell, Bertrand, 23
Rutgers University, 73
Rutherford, Ernest, 110, 130

Sagan, Carl, vii, xiii, 19, 224, 244, 247
Sakharov, Andrey, 135–136, 139, 142, 143, 199
Salam, Abdus, 121, 140
Salieri, Antonio, 29
Sandemanian Church, 50
Saturn, 24, 26, 157, 201
Scalar fields, 85–90, 94, 96, 97, 156
Schoenberg, Arnold, 102–103
Schopf, J. William, 5
Schrödinger, Erwin, xiv, 6, 111, 149
Scripps Research Institute, 184
Search for Extraterrestrial Intelligence (SETI), 246–247
Shaffer, Peter, 29
Shapiro, Robert, 183
Shaposhnikiv, Mikhail, 142–143, 145
Shelley, Mary, 13, 161–162
Shelley, Percy, 161
Siddharta Gautama, 20*n*
Sidereus Nuncius (*Starry Messenger*) (Galileo Galilei), 59
Single-celled organisms, 183, 209, 237
Smolin, Lee, 125
Smoot, George, 75
Soai, Kenso, 197
Socrates, 22, 128, 223
Sodium, 157
Sorcerer's Apprentice (Dukas), 196–197
Soudan 2 detector, 138
Souls, 163
Space-time, 67–68
Spallanzani, Lazzaro, 188
Spatial symmetries, 105
Special theory of relativity, 6, 15, 51–52, 67–68, 111, 113, 134
Speed of light, 51–52, 72, 80–81, 95, 229, 230
Speusippus, 23
Spielberg, Steven, 247
Spinoza, Baruch, 217
Spirituality, 18–19
Sponges, 183, 183*n*
Spontaneous generation, 187–190
Spores, 166
Standard Model of particle physics, 85–87, 93, 104, 123, 125, 134, 139, 140, 143, 145–146, 148, 225
Stapledon, Olaf, 234
Star Maker, The (Stapledon), 234

Star Trek, 245
Stars, 6, 64, 65*n*, 66, 72, 73, 91–94, 114, 156–157, 165, 212, 229–230, 242, 249
Steady-state model, 43
Stecker, Floyd, 115
Stoppard, Tom, 6
Strangeness, 118, 119
Stravinsky, Igor, 103
String landscape, 5
String theory. *See* Superstring theories
Stromatolites, 169
Strong nuclear force, 87, 96*n*, 117–121, 125, 127, 136, 139
Subatomic particles, 93, 112, 116
Sulfuric compounds, 165
Sun, 26, 27, 31–32, 37*n*, 54, 59, 72, 91, 92*n*, 94, 111, 131, 157, 179*n*, 238
Sun-centered model, 27–29, 31–32, 39, 59, 91
Super-Kamiokande detector, 138
Supermind, 60, 62, 106
Supernatural powers, 9–11, 17, 165, 189, 212, 216, 222
Supernaturalists, 216, 224
Supernovae, 95, 157, 235
Superstring theories, xiv, 69–70, 90, 97, 98, 125, 126, 147, 149, 225, 230–231
Supersymmetry (SUSY), 137–139, 146, 148
Susskind, Leonard, 96

Tartaric acid, 190–192
Tau, 121
Taylor, John G., 149
Tegmark, Max, 220
Telescope, invention of, 59
Tertiary era, 205, 208
Thales, 22, 23, 118, 123, 147, 149
Theory of Everything, xiv, 7, 218–219
Thermal equilibrium, 79, 136, 139
Thermodynamics, second law of, 225
Thomson, J. J., 109–110
Thorarinson, Joel, 202
Thorium, 110
Thymine, 209
Tides, 175, 175*n*
Time, asymmetry of, 43–44
Time-reversal invariant, 133–134
Time travel, 15
Titan, 201
Transverse wave, 190
Tritium, 55, 110
Trodden, Mark, 146
2001: A Space Odyssey (Clarke), 241
2001: A Space Odyssey (movie), 13, 244

UFOs (unidentified flying objects), 246
Ultraviolet radiation, 61, 62, 165, 166, 183, 185, 201, 204, 211
Unification theory, xiv, 22–23, 38, 50, 57, 69, 87, 88, 90, 97, 98, 123–129, 131, 140, 217–220, 222, 225, 230
  critique of, 147–154
Unified field theory, 7
Uniformitarianism, 206–208
Universal gravity, theory of, 82
University of Bristol, 196
University of Madrid, 202
University of Manchester, 185
Uranium, 64–65, 110
Uranus, 91, 157
Urey, Harold, 164–165
  Miller-Urey spark-of-life experiments, 164–166, 180, 185, 195, 196, 199, 236

Vaccines, 187
Vector fields, 85
Venus, 26, 59, 157, 236

Viedma, Cristóbal, 202
Vilenkin, Alex, 221
Visible eye, 54
Volcanism, 164, 205, 208
Volta, Alessandro, 160
Voltaic cell, 161
Voltaire, 60
Voyage of the Beagle (C. Darwin), 171

Wake Forest University, 202
Walker, Sara, 174, 202
War of the Worlds (Wells), 241
Ward, Peter, 238, 240
Water, 85, 141, 157, 164, 173, 174, 235, 236, 238
Watson, James, 196
Weak gauge bosons, 116, 121, 125, 127, 137, 140, 141, 144
Weak nuclear force, 45, 87, 96*n*, 121, 136, 139, 140, 141, 200, 201, 204
Weinberg, Steven, vii, 19, 45, 46, 55, 64, 80, 104, 121, 123, 126, 126*n*, 140, 163, 221
Wells, H. G., 241
Wheeler, John, 217
Wigner, Eugene, 226
Wilczek, Frank, 19, 120, 126, 126*n*
William of Ockham, 82
Wilson, Edward O., 224
Wilson, Robert, 43, 57, 64, 74
Woit, Peter, 125
Wolf Man, The (movie), 206

Xenocrates, 23

Yang, C. N., 132
Yukawa, Hideki, 118

Zeus, 20*n*
Zircon, 170
Zweig, George, 118

ABOUT THE AUTHOR

Marcelo Gleiser is the Appleton Professor of Natural Philosophy and professor of physics and astronomy at Dartmouth College. He is a Fellow of the American Physical Society and the recipient of the Presidential Fellow Award from the White House and the National Science Foundation, as well as several literary awards. His research focuses on questions of origins—of the universe, of matter, and of life—bringing together particle physics and cosmology, and, more recently, astrobiology. He is the author of *The Dancing Universe: From Creation Myths to the Big Bang* and *The Prophet and the Astronomer: A Scientific Journey to the End of Time*.